MW00674654

AUTOMATED EEG-BASED DIAGNOSIS OF NEUROLOGICAL DISORDERS

Inventing the Future of Neurology

Dedicated to

Dr. Nahid, Dr. Anahita, Dr. Amir, Mona, and Cyrus Dean
Adeli

Kanan, Somashree, Manisha, and Indrajit
Ghosh Dastidar

Preface

Based on the authors' ground-breaking research, the book presents an ideology, a novel multi-paradigm methodology, and advanced computational models for automated EEG-based diagnosis of neurological disorders that the authors believe will be the wave of the future and an important tool in the practice of neurology. It is based on adroit integration of three different computing technologies and problem solving paradigms: neural networks, wavelets, and chaos theory. The book also includes three introductory chapters in order to introduce the readers to these three different computing paradigms.

Epilepsy, the primary application focus of the book, is a common disorder affecting approximately 1% of the population in the United States and is commonly accompanied by intermittent abnormal firing of neurons in the brain leading to recurrent and spontaneous seizures (with no apparent external cause or trigger). At present, epileptic seizure detection and epilepsy diagnosis are performed primarily based on visual examinations of electroencephalograms (EEGs) by highly trained neurologists. While many attempts have been reported in the literature none has been accurate enough to perform better than practicing neurologists/epileptologists. Effective algorithms for automatic seizure detection and prediction can have a far-reaching impact on diagnosis and treatment of epilepsy.

In this book, the clinical epilepsy and seizure detection problem is modeled as a three-group classification problem. The three subject groups are: a) healthy subjects (normal EEG), b) epileptic subjects during a seizure-free interval (interictal EEG), and c) epileptic subjects during a seizure (ictal EEG).

Epilepsy diagnosis is modeled as the classification of normal EEGs and inter-ictal EEGs. Seizure detection is modeled as the classification of interictal and ictal EEGs.

The approach presented in this book challenges the assumption that the EEG represents the dynamics of the entire brain as a unified system and needs to be treated as a whole. On the contrary, an EEG is a signal that represents the effect of the superimposition of diverse processes in the brain. There is no good reason why the entire EEG should be more representative of brain dynamics than the individual frequency sub-bands. In fact, the sub-bands may yield more accurate information about constituent neuronal activities underlying the EEG and, consequently, certain changes in the EEGs that are not evident in the original full-spectrum EEG may be amplified when each sub-band is analyzed separately. This is a fundamental premise of the authors' approach.

After extensive research and discovery of mathematical markers, the authors present a methodology for epilepsy diagnosis and seizure detection with a high accuracy of 96%. The technology presented in the book outperforms practicing neurologists/epileptologists. It has the potential to impact and transform part of the neurology practice in a significant way.

The book also includes some preliminary results toward EEG-based diagnosis of Alzheimer's disease (AD) which is admittedly in its infancy. But the preliminary findings presented in the book provide the potential for a major breakthrough for diagnosis of AD. The methodology presented in the book is general and can be adapted and applied for diagnosis of other brain disorders. The senior author and his research associates are currently extending the work to automated EEG-based diagnosis of AD and other neurological disorders such Attention-Deficit/Hyperactivity Disorder (ADHD) and autism.

A second contribution of the book is presenting and advancing Spiking

Neural Networks as the seminal foundation of a more realistic and plausible third generation neural network. It is hoped the fundamental research in this area of neuronal modeling will advance in the coming years resulting in more powerful computational neural network models not only for diagnosis of neurological disorders but also many other complex and intractable dynamic pattern recognition and prediction phenomena.

Hojjat Adeli

Samanwoy Ghosh-Dastidar

Acknowledgments

Since the authors' overarching goal in this work was to make a significant impact on the future practice of neurology the neurological aspects of the models presented in the book were reviewed and corroborated by a board-certified neurologist, Nahid Dadmehr, M.D., in practice for nearly two decades. Her contribution to this work and to the future of neurology practice is gratefully acknowledged. We are also grateful to Dr. Dennis Duke of Florida State University and Dr. Kerry Coburn of Mercer University for providing EEG data on AD patients used in Chapter 12.

Parts of the work presented in this book were published by the authors in several research journals: *IEEE Transactions on Biomedical Engineering* (published by IEEE), *Neural Networks* (published by Elsevier), *Neuroscience Letters* (published by Elsevier), *Journal of Alzheimer's Disease* (published by IOS Press), *Clinical EEG and Neuroscience* (published by EEG and Clinical Neuroscience Society), *Integrated Computer-Aided Engineering* (published by IOS Press), and *International Journal of Neural Systems* (published by World Scientific), as noted in the list of references and cited throughout the book. Chapter 6 is based on a journal article by the senior author, his former research associate, Ziqin Zhou, and Dr. Nahid Dadmehr, and is reproduced by permission of Elsevier, the publisher of the journal.

About the Authors

Hojjat Adeli received his Ph.D. from Stanford University in 1976 at the age of 26 after graduating from the University of Tehran in 1973 with the highest rank among the graduates of the entire College of Engineering. He is currently Professor of Civil and Environmental Engineering and Geodetic Science and the holder of the Abba G. Lichtenstein Professorship at The Ohio State University. He is also Professor of Aerospace Engineering, Biomedical Engineering, Biomedical Informatics, Electrical and Computer Engineering, Neurological Surgery and Neuroscience by courtesy. He has authored over 450 research and scientific publications in various fields of computer science, engineering, applied mathematics, and medicine including 240 journal articles and fourteen books. His wide-ranging research has been published in 74 different journals. He has also edited thirteen books. He is the Founder and Editor-in-Chief of the international research journals *Computer-Aided Civil and Infrastructure Engineering*, in publication since 1986, and *Integrated Computer-Aided Engineering*, in publication since 1993, and Editor-in-Chief of *International Journal Neural Systems*. He is the quadruple winner of The Ohio State University College of Engineering Lumley Outstanding Research Award. In 1998 he received the *Distinguished Scholar Award*, The Ohio State University's highest research award *"in recognition of extraordinary accomplishment in research and scholarship"*. In 2005, he was elected Honorary/Distinguished Member, American Society of Civil Engineers:*"for wide-ranging, exceptional, and pioneering contributions to computing in civil engineering disciplines and extraordinary leadership in advancing the use of computing and information* *technologies in many engineering disciplines throughout the world"*. In 2007, he received The Ohio State University College of Engineering Peter L. and Clara M. Scott Award for Excellence in Engineering Education as well as the Charles E. MacQuigg Outstanding Teaching Award. He has presented Keynote Lectures at 73 conferences held in 40 different countries. He has been on the organizing or scientific committee of over 300 conferences held in 58 countries. He holds a U.S. patent for his neural dynamics model (with his former Ph.D. student, Prof. H.S. Park).

Samanwoy Ghosh-Dastidar received his B.E. from the Indian Institute of Technology, Roorkee, India in 1997, and M.S. and Ph.D. from The Ohio State University in 2002 and 2007, respectively. He is currently Principal Biomedical Engineer at ANSAR Medical Technologies, Inc. in Philadelphia, PA. He has published 14 articles in refereed journals. He is listed in Who's Who in America. His interests include neuroengineering; mathematical modeling of neural dynamics and mechanisms of neurological disorders; human behavior and cognitive neuroscience with specific application to modeling human interaction with the environment (including Brain-Computer Interfaces); data mining and signal processing of physiological signals (primarily, EEG, EKG, and HRV); pattern recognition models for discovering functional imaging (fMRI) and EEG correlates of brain function and dysfunction; neuromodulation and neurostimulation based disease treatment strategies; biologically inspired intelligent computational models for dynamic function and parameter estimation (artificial neural networks, genetic algorithms, and fuzzy logic); and data fusion.

List of Figures

List of Tables

Contents

III Automated EEG-Based Diagnosis of Alzheimer's Disease 183

Part I

Basic Concepts

1

Introduction

This book presents a novel approach for automated electroencephalogram (EEG)-based diagnosis of neurological disorders such as epilepsy based on the authors' ground-breaking research in the past six years. It is divided into four parts. Basic concepts necessary for understanding the book are reviewed briefly in Chapters 2, 3, and 4. Chapter 2 introduces the readers to time-frequency analysis and wavelet transform. Chaos theory is described in Chapter 3. Chapter 4 presents the design of different classifiers.

At present, epileptic seizure detection and epilepsy diagnosis are performed mostly manually based on visual examinations of EEGs by highly trained neurologists. In epilepsy monitoring units, seizure detection is performed by semi-automated computer models. However, such models require close human supervision due to frequent false alarms and missed detections. Epilepsy diagnosis is more complicated due to excessive myogenic artifacts, interference, overlapping symptomatology with other neurological disorders, and low understanding of the precise mechanism responsible for epilepsy and seizure propagation.

While many attempts have been reported in the literature none is accurate enough to perform better than a practicing neurologist. In this book, the clinical epilepsy and seizure detection problem is modeled as a three-group classification problem. The three subject groups are: a) healthy subjects (normal EEG), b) epileptic subjects during a seizure-free interval (interictal EEG), and c) epileptic subjects during a seizure (ictal EEG). Epilepsy diagnosis is modeled as the classification of normal EEGs and interictal EEGs. Seizure detection is modeled as the classification of interictal and ictal EEGs. Part II is devoted to automated EEG-based diagnosis of epilepsy starting with an introduction to EEG and epilepsy in Chapter 5. Chapter 6 presents analysis of EEGs in an epileptic patient using wavelet transform. Chapter 7 presents a wavelet-chaos methodology for analysis of EEGs and EEG sub-bands. Chapter 8 describes a mixed-band wavelet-chaos neural network methodology for classifying EEGs obtained from the three subject groups. Chapter 9 shows how the methodology can be further improved by employing a principal component analysis (PCA)-enhanced cosine radial basis function (RBF) neural network.

While EEG-based diagnosis of Alzheimer's disease (AD) is still in its infancy, the authors' preliminary findings provide the potential for a major breakthrough for diagnosis of AD. Part III deals with AD. Chapter 10 presents a review of imaging, classification, and neural computational models for AD. Chapter 11 reviews analyses of EEGs obtained from AD patients. Chapter 12 presents a spatio-temporal wavelet-chaos methodology for EEG-based diagnosis of AD with some preliminary results.

Part IV is devoted to a new and advanced concept, Spiking Neural Networks (SNN), referred to as the third generation neural networks. Spiking neurons, their biological foundations, and training algorithms are presented in Chapter 13. Chapter 14 presents an improved SNN and its application to EEG classification and epilepsy diagnosis and seizure detection. Chapter 15 describes a new supervised learning algorithm for multi-spiking neural networks (MuSpiNN). Chapter 16 presents applications of MuSpiNN to EEG classification and epilepsy diagnosis and seizure detection. Finally, some future directions are noted in Chapter 17.

2

Time-Frequency Analysis: Wavelet Transforms

2.1 Signal Digitization and Sampling Rate

A time series is defined as a series of discrete data points representing measurements of some physical quantity over time. Following this definition, any signal can be conceptualized as a time series (provided that the dependent variable is time). However, since most real world signals are analog and continuous, they have to be digitized or discretized to fit the definition of a time series. To properly differentiate between the two, a continuous signal is denoted by $f(t)$ and the corresponding discretized time series is denoted by $f[n]$, where n is the sample number ($n \in \mathbb{Z}$, where $\mathbb{Z}$ is the set of integers). The difference between the two is illustrated in Fig. 2.1, which shows the continuous signal $f(t) = sin(t)$ plotted as a function of time and the corresponding discretized time series $f[n] = sin(nT)$, where T is the sampling interval, sampled at every one-second interval ($T = 1$ second). Since all signals considered in this book are time-dependent and digitized, the terms *signal* and *time series* will be used interchangeably unless specified otherwise.

An important concept in signal analysis is that of the sampling rate (or the sampling frequency) which defines the number of samples or measurements taken per second from the continuous signal to generate the discretized sig-

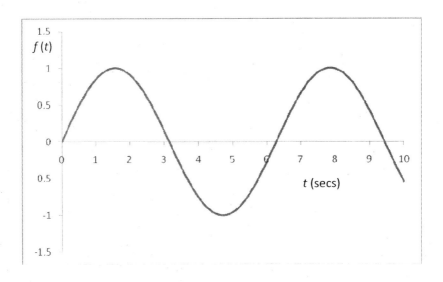

(a)

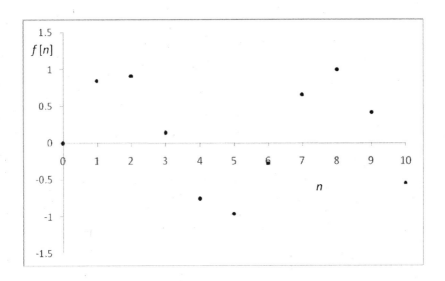

(b)

FIGURE 2.1
(a) A continuous signal $f(t) = sin(t)$ plotted as a function of time and (b)
the discretized form of the same signal sampled at every 1 second interval
(sampling frequency = 1 Hz)

nal. Numerically, the sampling rate is computed as the inverse of the sampling interval. For instance, in Fig. 2.1(b), the sampling rate is 1 data point per second (sampling frequency = 1 Hz). The significance of the sampling frequency is illustrated in Fig. 2.2. Figure 2.2 shows the same continuous signal in Fig. 2.1(a) but with an additional blip between $t = 1$ and 2 seconds and its discretized form. However, despite this difference in the signals, it can be clearly observed in Fig. 2.2(b) that the discretized time series is identical to that in Fig. 2.1(b). This implies that enough samples were not selected from the signal shown in Fig. 2.2(a) for accurate characterization of the sharp blip in the signal. In general, if a signal contains transient waveforms or high-frequency components, it must be sampled at a higher rate.

It is clear that the selection of the sampling rate is very important for accurate signal representation. According to the Nyquist-Shannon sampling theorem in information theory, a signal can be completely characterized if it is sampled at a rate that is greater than twice the highest frequency contained in the signal. In other words, if the highest frequency component in a signal has a frequency of f_{Max} Hz, then the signal must be sampled at $2f_{\text{Max}}$ Hz to prevent information loss (similar to that demonstrated in Fig. 2.2). A corollary of this theorem is that if the signal is sampled at $2f_{\text{Nyq}}$ Hz, then the maximum useful frequency contained in the signal is limited to f_{Nyq}, which is called the *Nyquist Frequency*.

2.2   Time and Frequency Domain Analyses

Time-domain analysis of any signal primarily involves an analysis of the measurements or data points in time as they appear as part of the waveform. As a result, time-domain measures are often the easiest to visualize. In time-domain

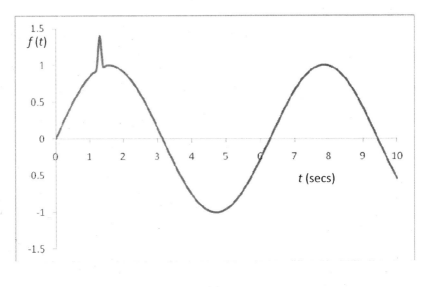

(a)

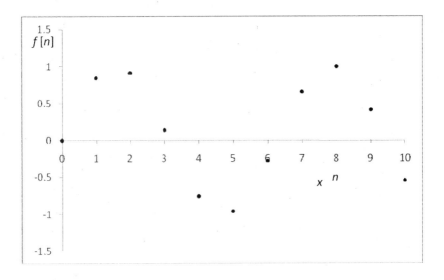

(b)

FIGURE 2.2
(a) A continuous signal $f(t) = sin(t)$ with an additional blip between $t = 1$
and $t = 2$ seconds plotted as a function of time and (b) the discretized form
of the same signal sampled at every 1 second interval (sampling frequency =
1 Hz)

analysis, the numerical values of the measurements recorded at a series of instants of times are of paramount importance. The most common time-domain signal characteristics include statistical measures such as the mean, median, and standard deviation. More advanced statistical measures are also employed to better understand the temporal characteristics of the signal or how the signal changes with time. Such measures include the change in signal, the rate of change of the signal, moving average, autocorrelation, and autoregression.

Although time-dependent signals are easy to visualize as a function of time, certain characteristics of the signal such as component frequencies of the signal cannot be obtained from a visual inspection. Moreover, signal manipulations and computations such as convolutions and filtering are often much simpler to perform in the frequency domain. This is not surprising because many of these concepts are frequency domain concepts. As a result, many of the modern signal processing techniques are based on frequency-domain measures. Some signal processing techniques such as smoothing can be performed in the time or frequency domain but the methods used may be different for each domain.

The Fourier transform has long been a staple of the signal processing field and is used to express any signal in the time domain into the frequency domain. The basic assumption is that any signal can be expressed as the sum of a number of sinusoids with varying frequencies and amplitudes. This is further extended to be applicable to real-world signals, which are usually not perfectly periodic. Therefore, the Fourier transform decomposes the given signal into its constituent frequency components. Assuming that both time and frequency domain representations of the signal are continuous, the Fourier transform of the continuous signal $f(t)$ is expressed as:

$$F(\omega) = \int_{-\infty}^{\infty} f(t)e^{-i\omega t}dt, \qquad \omega \in \mathbb{R} \tag{2.1}$$

where $\mathbb{R}$ is the set of real numbers. The signal can be reconstructed using the inverse Fourier transform as:

$$f(t) = \frac{1}{2\pi} \int_{-\infty}^{\infty} F(\omega)e^{i\omega t}d\omega, \qquad t \in \mathbb{R} \tag{2.2}$$

The discrete-time Fourier transform (DTFT) is used to compute the Fourier transform of the discretized time series $f[n]$ as:

$$F(\omega) = \sum_{-\infty}^{\infty} f[n]e^{-i\omega n}, \qquad \omega \in \mathbb{R} \tag{2.3}$$

In this case, the time-domain representation is discrete but the frequency domain representation remains continuous. The time series can be reconstructed as:

$$f[n] = \frac{1}{2\pi} \int_{-\pi}^{\pi} F(\omega)e^{i\omega n}d\omega, \qquad n \in \mathbb{Z} \tag{2.4}$$

The discrete Fourier transform (DFT) is the discretized form of the DTFT evaluated at $\omega = 2\pi k/N$ and is expressed as:

$$F[k] = \sum_{n=0}^{N-1} f[n]e^{-i\frac{2\pi}{N}kn}, \qquad k = 0, 1, \ldots, N-1 \tag{2.5}$$

where N is the number of samples in time series $f[n]$. In this case, both time and frequency domain representations of the signal are discrete. The time series can be reconstructed as:

$$f[n] = \frac{1}{N} \sum_{n=0}^{N-1} F[k]e^{i\frac{2\pi}{N}kn}, \qquad n \in \mathbb{Z} \tag{2.6}$$

The most commonly used algorithm for computing the DFT is a computationally optimized algorithm known as the fast Fourier transform (FFT).

2.3 Time-Frequency Analysis

2.3.1 Short Time Fourier Transform (STFT)

Real world signals are usually non-stationary and contain many transient events. The infinite basis functions used in Fourier analysis are primarily suitable for extracting frequency information from periodic, non-transient signals. Therefore, Fourier transforms are unable to appropriately capture the transient features in a signal. Moreover, Fourier coefficients of a signal are determined from the entire signal support. This implies that the frequency spectrum of a signal as a result of the Fourier transform is not localized in time and the temporal information cannot be extracted readily from the Fourier transform coefficients. Consequently, if additional data are added over time, the Fourier transform coefficients change. Any localized event in a signal cannot be easily located in time from its Fourier transform.

The retention of only the frequency information of a signal and not the time information is a major disadvantage. The short-time Fourier transform (STFT) attempts to overcome this shortcoming by mapping a signal into a two dimensional function of frequency and time (Gabor, 1946). STFT is a time-frequency analysis method in which time and frequency information is localized by a uniform-time sliding window for all frequency ranges. In other words, only a small local window of the signal is analyzed using Fourier transform. This window is then shifted along the signal to analyze the next signal segment, and so on until the entire signal is analyzed.

Many window functions are available for use with STFT. A rectangular window gives equal weights to all data points within the window. A special case of the STFT is the Gabor transform (Gabor, 1946) in which a Gaussian function is employed as the window and the center of the function is located on the time instant being analyzed. Using a Gaussian window means that points within the window that are closest to the center of the window have a greater weight as compared to points in the periphery. However, the precision of this method is limited by the window size, which stays the same for all frequencies.

2.3.2 Wavelet Transform

In the last two decades, many models have been developed that are based on the wavelet transform. Wavelets can be literally defined as small waves that have limited duration and zero average values. An example wavelet is compared to a sine wave in Fig. 2.3. They are mathematical functions capable of localizing a function or a set of data in both time and frequency. Wavelets can be stretched or compressed and used to analyze the signal at various levels of resolution. Stretched and compressed versions of the wavelet in Fig. 2.3 are shown in Fig. 2.4. Thus, the wavelet transform acts like a *mathematical microscope*, zooming into small scales to reveal compactly spaced events in time and zooming out into large scales to exhibit the global waveform patterns.

The root of wavelets can be traced back to the thesis of Haar in 1909 (Daubechies, 1992). In the 1930s, scale-varying basis functions were developed in mathematics, physics, and electronics engineering, as well as in seismology independently. The broad concept of wavelets was introduced in the mid 1980s by Grossman and Morlet (1984). The wavelet transform is an effective tool in signal processing due to attractive properties such as time-frequency localization (obtaining features at particular times and frequencies), scale-space analysis (extracting features at various locations in space at different scales) and multi-rate filtering (separating signals with varying frequency content) (Mallat, 1989; Daubechies, 1992; Meyer, 1993; Jameson et al., 1996; Burrus

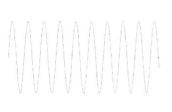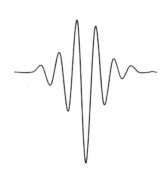

FIGURE 2.3
General forms of the sine function and a generic wavelet [Adapted from Hubbard (1998)]

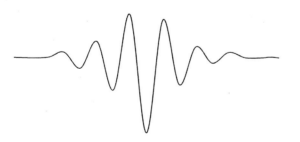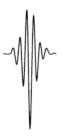

FIGURE 2.4
A stretched (large scale) and compressed (small scale) wavelet [Adapted from Hubbard (1998)]

et al., 1998; Mallat, 1998; Rao and Bopardikar, 1998; Adeli and Samant, 2000; Samant and Adeli, 2000; Ghosh-Dastidar and Adeli, 2003).

Similar to Fourier transform, wavelet transform can be discrete or continuous. The continuous wavelet transform (CWT) can operate at every scale, from the scale of the original signal to a maximum, which is determined on the basis of need and available computational power. In CWT, the signal to be analyzed is matched and convolved with the wavelet basis function at continuous time and frequency increments. However, it must be remembered that even in CWT the signal itself has to be digitized. Continuous time and frequency increments merely indicate that data at every digitized point or increment is used. The CWT is also continuous in the sense that during analysis, the wavelet is shifted smoothly over the full domain of the function being transformed.

On the other hand, the discrete wavelet transform (DWT) uses dyadic (powers of 2) scales and positions (based on powers of 2) making it computationally very efficient without compromising accuracy (Karim and Adeli, 2002b; Adeli and Ghosh-Dastidar, 2004; Jiang and Adeli, 2005a,b). As a result, the original signal is expressed as a weighted integral of the continuous wavelet basis function. To avoid redundancy between the basis functions in DWT, these basis functions are often designed to be orthogonal, i.e., the inner product of any pair of basis functions is zero. In DWT, the inner product of the original signal with the wavelet basis function is taken at discrete points (usually dyadic to ensure orthogonality) and the result is a weighted sum of a series of basis functions.

The basis for wavelet transform is the wavelet function. Wavelet functions are families of functions satisfying prescribed conditions, such as continuity, zero mean amplitude, and finite or near finite duration, and orthogonality. This chapter covers a brief introduction to wavelet transforms. For a more detailed description, please refer to the literature (Chui, 1992; Strang, 1996; Samant and Adeli, 2000; Karim and Adeli, 2002b).

If the signal $f(t)$ is a square integrable function of time t, i.e., $\sum_{-\infty}^{\infty} |f(t)|^2 dt$ is finite, then the continuous wavelet transform of $f(t)$ is defined as (Chui, 1992):

$$W_{a,b} = \int_{-\infty}^{\infty} f(t) \frac{1}{\sqrt{|a|}} \psi^* \left(\frac{t-b}{a} \right) dt \qquad (2.7)$$

where $a, b \in \mathbb{R}, a \neq 0$, and the asterisk (*) denotes the complex conjugate of the function. The wavelet function is defined as:

$$\psi_{a,b}(t) = \frac{1}{\sqrt{|a|}} \psi^* \left(\frac{t-b}{a} \right) \qquad (2.8)$$

where the factor $1/\sqrt{|a|}$ is used to normalize the energy so that the energy stays at the same level for different values of a and b, i.e., at different levels of resolution. Equation (2.7) can now be expressed in the general form of the CWT as:

$$W_{a,b} = \int_{-\infty}^{\infty} f(t) \psi_{a,b}^*(t) dt \qquad (2.9)$$

The wavelet function $\psi_{a,b}(t)$ becomes narrower when a is increased and displaced in time when b is varied. Therefore, a is called the scaling parameter which captures the local frequency content and b is called the translation parameter which localizes the wavelet basis function in the neighborhood of time $t = b$. The inverse wavelet transform is used to reconstruct the signal as:

$$f(t) = \int_{-\infty}^{\infty} \int_{-\infty}^{\infty} W_{a,b} \psi_{a,b}(t) da \, db \qquad (2.10)$$

Analyzing the signal using CWT at every possible scale a and translation

b is computationally very intensive. Therefore, to make the computational burden manageable, discrete wavelets are obtained in DWT by considering discrete values of the scaling and translation parameters a and b based on powers of two (dyadic scales and translations):

$$a_j = 2^j, \qquad b_{j,k} = k2^j \qquad \text{for all } j, k \in \mathbb{Z} \tag{2.11}$$

Substituting in Eq. (2.8) yields:

$$\psi_{j,k}(t) = 2^{-j/2}\psi(2^{-jt} - k) \qquad \text{for all } j, k \in \mathbb{Z} \tag{2.12}$$

The set of functions $\psi_{j,k}(t)$ forms a basis for the square integrable space $L^2(\mathbb{R})$. The set of basis functions $\psi_{j,k}(t)$ is selected to be orthogonal. As a result, the redundant information in CWT is discarded and the original signal can be reconstructed from the resulting wavelet coefficients accurately and efficiently without any loss of information (Strang, 1996) as:

$$f(t) = \sum_{j=-\infty}^{\infty} \sum_{k=-\infty}^{\infty} W_{j,k}\psi_{j,k}(t) \tag{2.13}$$

It is clear that an infinite number of wavelets would be required to define the original signal. To make the number of wavelets finite, the concepts of scaling function and multi-resolution analysis need to be introduced (Daubechies, 1992; Burrus et al., 1998; Goswami and Chan, 1999). In essence, a finite number of wavelets are employed to represent the signal down to a certain scale and the remainder of the signal is represented by the scaling function. Similar to the original signal, the scaling function may be expressed as:

$$\varphi(t) = \sum_{j=-\infty}^{\infty} \sum_{k=-\infty}^{\infty} W_{j,k}\psi_{j,k}(t) \tag{2.14}$$

The general representation of the scaling function (Eq. 2.14) is the same as that of the signal (Eq. 2.14) because the scaling function can be conceptualized as a low-pass filtered version of the signal. The only difference is that the scaling function employs an infinite number of wavelets up to a certain scale j as described shortly from the perspective of multi-resolution analysis.

Multi-resolution analysis provides an effective way of implementing DWT (Mallat, 1989) in which the square integrable space $L^2(\mathbb{R})$ is decomposed into a direct sum of the subspaces W_j, where j lies in the range $-\infty$ to ∞. The square integrable space is expressed as:

$$L^2(\mathbb{R}) = \ldots W_{-3} \oplus W_{-2} \oplus W_{-1} \oplus W_0 \oplus W_1 \oplus W_2 \oplus W_3 \oplus \ldots \qquad (2.15)$$

This implies that at the boundary condition $j = -\infty$, the subspace covers only the lowest resolution signal which is the null set $\{\phi\}$ and at $j = \infty$, the subspace covers the original signal space $L^2(\mathbb{R})$ with every detail. If the closed subspaces V_j are defined as:

$$V_j = W_{j+1} \oplus W_{j+2} \oplus W_{j+3} \oplus \ldots \qquad \text{for all } j \in \mathbb{Z} \qquad (2.16)$$

where $\oplus$ indicates direct sum, then the subspaces V_j are a multi-resolution approximation of the square integrable space $L^2(\mathbb{R})$. Thus, the subspaces W_j are the orthogonal complement of the subspaces V_j:

$$V_{j-1} = V_j \oplus W_j \qquad \text{for all } j \in \mathbb{Z} \qquad (2.17)$$

For the DWT, the scaling and wavelet functions are expressed as (Burrus et al., 1998; Goswami and Chan, 1999):

$$\varphi(2^j t) = \sum_{k=-\infty}^{\infty} h_0(k) \sqrt{2} \varphi(2^{j+1} t - k) \qquad (2.18)$$

and

$$\psi(2^j t) = \sum_{k=-\infty}^{\infty} h_1(k)\sqrt{2}\varphi(2^{j+1}t - k) \qquad (2.19)$$

where $h_0(k)$ represents the sequence of scaling function coefficients or the scaling filter and $h_1(k)$ represents the sequence of wavelet function coefficients or the wavelet filter.

Finally, the multi-resolution decomposition formula for DWT of the original function $f(t)$ is obtained as:

$$f(t) = \sum_{k=-\infty}^{\infty} c_k \varphi_k(t) + \sum_{k=-\infty}^{\infty} \sum_{j=0}^{\infty} d_{j,k} \psi_{j,k}(t) \qquad (2.20)$$

where $d_{j,k}$ are the wavelet coefficients, and c_k are the scaling coefficients. In the right-hand side of Eq. (2.20), the first term represents an approximation of the general trend of the original signal and the second term represents the local details in the original signal. The wavelet coefficients $d_{j,k}$ multiplied by the dilated and translated wavelet function can be interpreted as the local residual error between successive signal approximations at scales $j-1$ and j. Therefore, the detail signal at scale j is computed as:

$$r_j(t) = \sum_{k=-\infty}^{\infty} d_{j,k} \psi_{j,k}(t) \qquad (2.21)$$

An efficient DWT filter algorithm can be implemented based on multi-resolution analysis. The high-pass filter $h_1(k)$ corresponding to the wavelet function $\psi_{j,k}(t)$ extracts the signal details and the low-pass filter $h_0(k)$ corresponding to the scaling function $\varphi_{j,k}(t)$ extracts the trend or coarser information in the signal. In the DWT filter implementation, the signal is first down-sampled (reduced by half). Next, the down-sampled signal is convolved with the high-pass filter to produce the detail wavelet coefficients and with the low-pass filter to produce the shape approximating scaling coefficients.

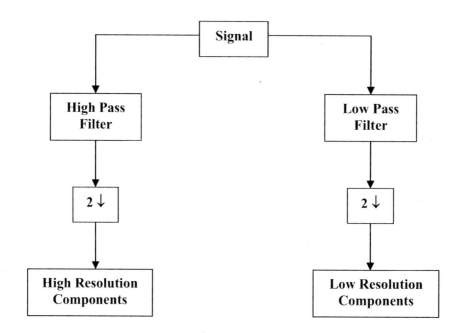

FIGURE 2.5
Level 1 wavelet decomposition

The purpose of the down-sampling is to avoid doubling the size of the sample during convolution.

After a single level decomposition, two sequences corresponding to the wavelet and scaling coefficients are obtained, as shown in Fig. 2.5. These sequences represent the high and low resolution components of the signal. However, such a single level decomposition may not always separate out all the desirable features. Therefore, the low resolution components are further decomposed into low and high resolution components after a second level of decomposition (as shown in Fig. 2.6), and so on. This process forms an integral part of the multi-resolution analysis where the decomposition process is iterated, with successive approximations being decomposed in turn, so that a single original signal may be examined at different levels with multiple resolutions. Any of the components can be used to reconstruct complete or filtered versions of the original signal, as shown in Fig. 2.7.

The following formula can be used when frequency information is needed instead of the scales (Abry, 1997):

$$F_a = \frac{F_c}{Ta} \tag{2.22}$$

where F_a is the pseudo-frequency in Hz corresponding to scale a, T is the sampling period, and F_c is the center frequency or dominant frequency of the wavelet in Hz, defined as the frequency with the highest amplitude in the Fourier transform of the wavelet function. Third order Daubechies wavelet and a sine function with the same frequency as the center frequency of the wavelet (equal to 0.8 Hz) are shown in Fig. 2.8. Fourth order Daubechies wavelet and a sine function with the same frequency as the center frequency of the wavelet (equal to 0.71 Hz) are shown in Fig. 2.9.

2.4   Types of Wavelets

The most appropriate type of wavelet to be used for a signal depends on the type of the signal. Haar's wavelet is the simplest wavelet that uses square wave functions, which can be translated and scaled to span the entire signal domain.

The Daubechies family of wavelets (Daubechies, 1988, 1992) is one of the most commonly used wavelets satisfying the orthogonality conditions, thus allowing reconstruction of the original signal from the wavelet coefficients. The Daubechies wavelet system, a higher order generalization of Haar's wavelet, was shown to have a superior smoothing effect on signals (Adeli and Samant,

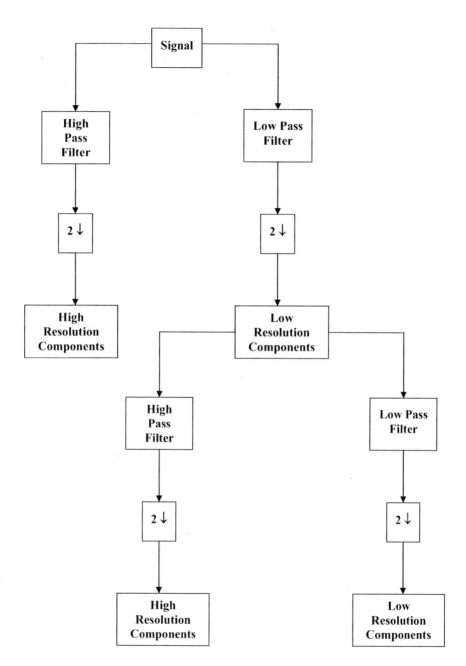

FIGURE 2.6
Level 2 wavelet decomposition

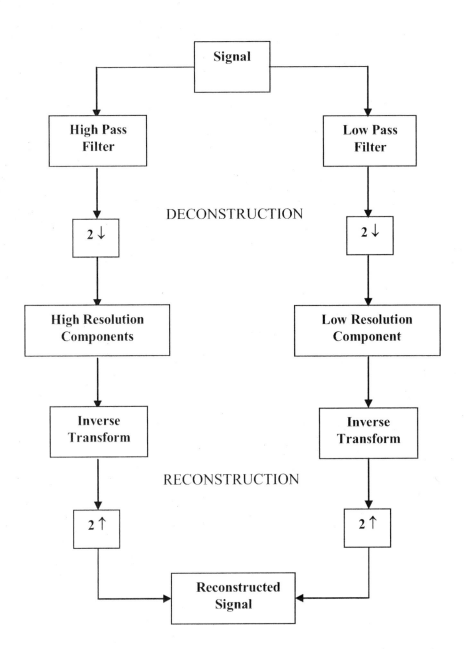

FIGURE 2.7
Signal reconstruction after level 1 wavelet decomposition

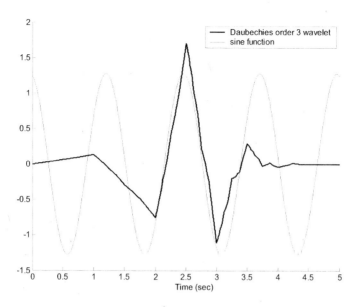

FIGURE 2.8
Third order Daubechies wavelet (center frequency: 0.8 Hz, center period: 1.25 sec) and a sine function (0.8 Hz)

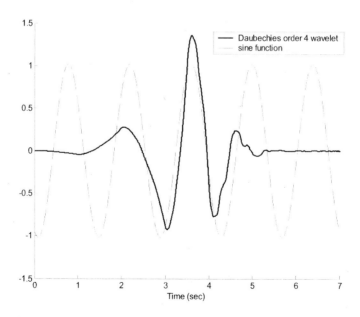

FIGURE 2.9
Fourth order Daubechies wavelet (center frequency: 0.71 Hz, center period: 1.4 sec) and a sine function (0.71 Hz)

2000; Samant and Adeli, 2000; Karim and Adeli, 2002a). Examples of wavelet and scaling functions for the Daubechies family of orthogonal wavelets are shown in Fig. 2.10. The Daubechies wavelet family is designed with maximum regularity (or smoothness). The first order Daubechies wavelet is actually the Haar wavelet. Daubechies wavelets are designed to have $N/2$ vanishing moments where N is the number of wavelet coefficients.

Daubechies also proposed the Coifman wavelets or *Coiflets*, which are more symmetric than the Daubechies wavelets. They are designed to have vanishing moment conditions for both the wavelet as well as the scaling functions (Daubechies, 1992; Burrus et al., 1998). The scaling functions of Coiflets have $N/3-1$ vanishing moments whereas the wavelet functions have $N/3$ vanishing moments. The filter coefficients for fourth order Coiflets are calculated using the quadrature mirror filter approach (Wickerhauser, 1994), which computes the coefficients of one filter as the mirror image of the other.

Harmonic wavelet transform is designed to achieve exact band separation in the frequency domain. The 0th order harmonic wavelet is a complex wavelet defined as the inverse Fourier transform of the following step function:

$$W(\omega) = \begin{cases} \dfrac{1}{2\pi} & \text{for } 2\pi \leq \omega \leq 4\pi \\ 0 & \text{otherwise} \end{cases} \tag{2.23}$$

Therefore, the 0th order harmonic wavelet function can be written as

$$w(x) = \frac{e^{i4\pi x} - e^{i2\pi x}}{i2\pi x} \tag{2.24}$$

Consider a band-limited step function in the frequency domain at scale j and translated by k steps of size $1/2^j$ expressed in the following form:

$$W(\omega) = \begin{cases} \dfrac{1}{2\pi} 2^{-j} e^{-i\omega k/2^j} & \text{for } 2\pi 2^j \leq \omega \leq 4\pi 2^j \\ 0 & \text{otherwise} \end{cases} \tag{2.25}$$

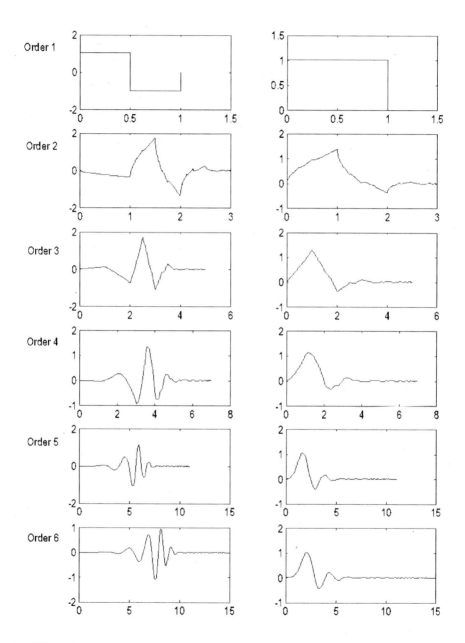

FIGURE 2.10
Daubechies wavelet and scaling functions of different orders

It can be shown that the discrete harmonic wavelet at a general level of decomposition j is the inverse Fourier transform of Eq. (2.25) as follows (Newland, 1993):

$$w(2^j x - k) = \frac{e^{i4\pi(2^j x - k)} - e^{i2\pi(2^j x - k)}}{i2\pi(2^j x - k)} \qquad (2.26)$$

The result of the discrete harmonic wavelet transform can be represented by a series of complex valued wavelet coefficients. The moduli of these complex wavelet coefficients represent the energy of the original signal at different frequency bands or decomposition levels appearing at different times. By investigating these complex wavelet coefficient moduli, the time-frequency characteristics of the original signal can be obtained.

2.5   Advantages of the Wavelet Transform

Wavelet transform is particularly effective for representing various aspects of signals such as trends, discontinuities, and repeated patterns especially in the analysis of non-stationary signals. Wavelet-based time-frequency decomposition of the signal can be used in object detection, feature extraction, and time-scale or space-scale analysis. Wavelet transform uses a variable window size over the length of the signal, which allows the wavelet to be stretched or compressed depending on the frequency of the signal (Mallat, 1989; Samant and Adeli, 2001; Zhou and Adeli, 2003; Jiang and Adeli, 2004). In other words, wavelet transform adapts the window size according to the frequency. At high frequencies, shorter windows are used (fine resolution) and at low frequencies, long windows are used (coarse resolution) to encompass the frequency content.

When wavelet transform is used to decompose a signal, the wavelet acts as its own window at each scale. As expected, the time resolution improves as analysis scale decreases. Wavelet transform is powerful for analyzing transient signals because both frequency (scale) and time information can be obtained simultaneously. Larger time intervals (corresponding to smaller values of a) are used for more precise extraction of low frequency information and shorter time intervals (corresponding to larger values of a) for the precise time localization of high-frequency information. Furthermore, if the wavelet basis function, Eq. (2.8), has a finite duration, then the frequency information obtained from the wavelet transform is localized in time. This results in excellent feature extraction even from non-stationary signals with transient waveforms and high frequency content (Petrosian et al., 1996, 2000b; Adeli et al., 2003, 2007; Ghosh-Dastidar and Adeli, 2003; Adeli and Ghosh-Dastidar, 2004). Therefore, for transient waveforms, wavelet analysis is superior to the Fourier transform and is used for the models presented in this book.

3

Chaos Theory

3.1   Introduction

The concept of chaos, as applied to nonlinear systems, is an exciting research topic of recent interest especially from the perspective of physiological signals. Almost all natural or real-world systems are nonlinear in nature and evolve with respect to some parameter such as time or space. As described in the previous chapter, systems that evolve with respect to time can be represented by a time series. Nonlinear systems often display behavior which *looks* random but, in fact, may be attributed to deterministic chaos. Figure 3.1 shows such a time series which, on the surface, looks random. In reality, the time series is generated by the recursive quadratic equation $x_{n+1} = rx_n(1 - x_n)$, known as the *logistic map*.

Deterministic chaos differs from random behavior in that it follows a set of specific rules, sometimes governed by simple differential equations, which determine the nonlinearity of the system. The predictability of the system depends upon the identification or approximation of the governing rules. Additionally, chaos in a system develops without external influence. A very small initial perturbation in the system results in a completely different scenario from the one in which there is no such perturbation (Williams, 1997; Smith, 1998). An example of this is shown in Fig. 3.2 where the logistic map (the

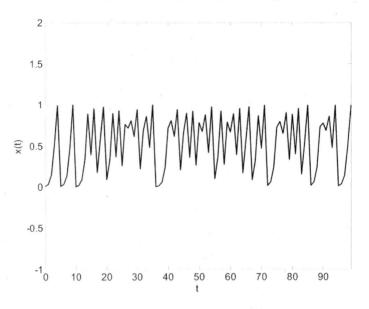

FIGURE 3.1

A chaotic time series generated using the logistic map $x_{n+1} = rx_n(1 - x_n)$
with the parameter r selected as 4

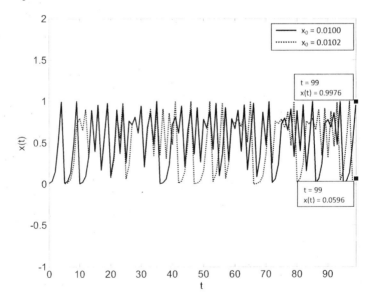

FIGURE 3.2

The logistic map with the parameter $r = 4.0$ is plotted for the initial conditions
$x_0 = 0.0100$ and $x_0 = 0.0102$ over a period of 100 time steps to demonstrate
the effect of a small change in initial conditions on the evolution of the system

parameter r is selected as 4.0) is plotted for two slightly different initial conditions, $x_0 = 0.0100$ and $x_0 = 0.0102$. The two cases evolve to very different states (0.9976 and 0.0596, respectively) after 100 time steps, as shown in Fig. 3.2.

Signal analysis and processing deals mainly with extracting relevant features from these real-world signals. In most cases, high frequency random fluctuations are discarded as noise. Even though the existence of nonlinearity is not sufficient proof for the existence of chaos, it is essential to analyze the signal and the noise for chaotic properties because discarding chaos in an evolving system may lead to an inaccurate analysis.

3.2   Attractors in Chaotic Systems

Nonlinear systems tend to gravitate toward specific regions in phase space known as attractors. Attractors can be of different types - point, limit cycle, toroidal, and chaotic (sometimes refered to as *strange*) depending on the behavior and state of the system. In general, the values of various parameters of the governing rules underlying deterministic chaos determine the shape of the attractor. In other words, a system is usually not chaotic for all parameter values. Incrementing the value of the parameters (in the positive or negative direction) could move a system from its initial non-chaotic attractor to a chaotic one. This transition usually shows characteristics such as period-doubling, intermittency, and quasi-periodicity.

The period-doubling characteristic, seen often in chaotic systems, is shown in the bifurcation diagram for the logistic map in Fig. 3.3. The bifurcation diagram represents the number of possible final states of x and their magnitudes for different values of the parameter r. For instance, if r is in the range 1 to

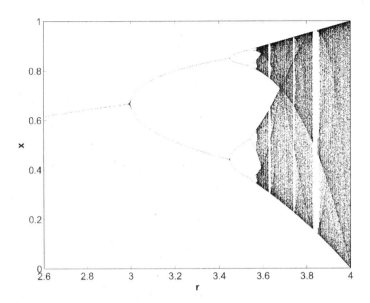

FIGURE 3.3
Bifurcation diagram for the logistic map showing the possible final outcomes
with changes in the value of r

3, x will converge to one final state (oscillations around this state are possible
prior to convergence for some values of r). These states are shown for $r = 2.0$
in Fig. 3.4 and $r = 2.9$ in Fig. 3.5. If r is in the range 3 to approximately
3.45, x will oscillate indefinitely between two possible final states (as shown
in Fig. 3.6 for $r = 3.3$). If r is approximately in the range 3.45 to 3.54, x will
oscillate indefinitely between four possible final states (as shown in Fig. 3.7 for
$r = 3.5$). The number of oscillations keeps doubling until around 3.57, beyond
which the attractor becomes chaotic. Figure 3.8 shows intermittency in the
evolution of the logistic map ($r = 3.8284$) where the attractor becomes chaotic
at certain times ($t = 220$ to 310, approximately in the plot) and non-chaotic
at others.

The Lorenz system consisting of three ordinary differential equations has
been extensively investigated and is probably the most commonly cited chaotic

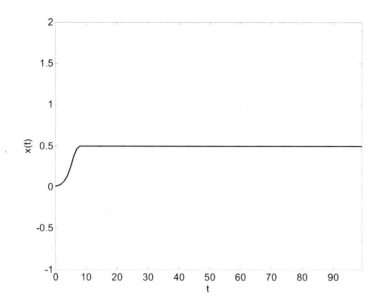

FIGURE 3.4
Logistic map for $r = 2$ showing convergence to one final state

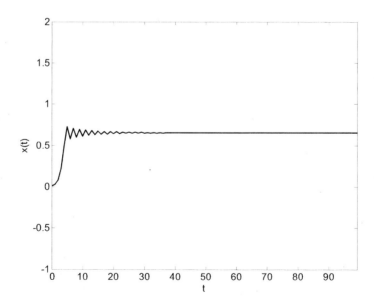

FIGURE 3.5
Logistic map for $r = 2.9$ showing some oscillation prior to convergence to one final state

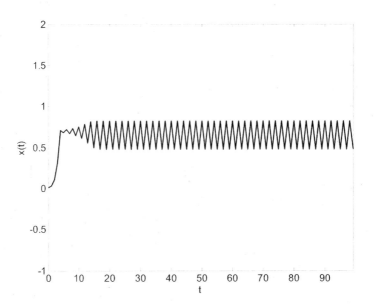

FIGURE 3.6
Logistic map for $r = 3.3$ showing indefinite oscillations between two final states

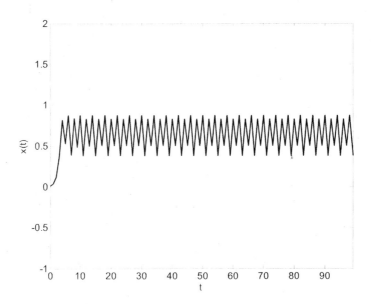

FIGURE 3.7
Logistic map for $r = 3.5$ showing indefinite oscillations between four final states

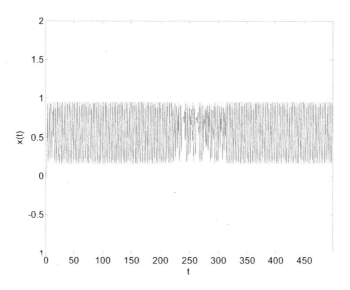

FIGURE 3.8

The logistic map for $r = 3.8284$ showing intermittency in the evolution of the system. The attractor becomes chaotic at certain times ($t = 220$ to 310, approximately) and non-chaotic at others.

system in the literature. The solution of the three differential equations with different parameter values yields different attractor shapes such as a point attractor and a chaotic attractor, as shown in Figs. 3.9 and 3.10, respectively. The characteristic butterfly-shaped attractor, also known as the Lorenz attractor, can be clearly observed in Fig. 3.10. To further illustrate the possible diversity in the attractor shape, two other chaotic attractors - the Hénon and Rössler attractors - are shown in Figs. 3.11 and 3.12, respectively.

Attractors are graphically represented using phase space plots of the nonlinear system. One of the most common representations is the lagged phase space plot where a signal is plotted against a lagged (or delayed) version of the same signal to highlight patterns in the temporal evolution of the system. The lagged phase space, using a lag of one time instant, for the logistic map is shown in Fig. 3.13. The graph is a plot of x_{n+1} versus x_n and highlights a

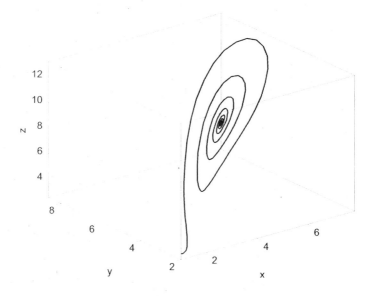

FIGURE 3.9
A point attractor for the Lorenz system

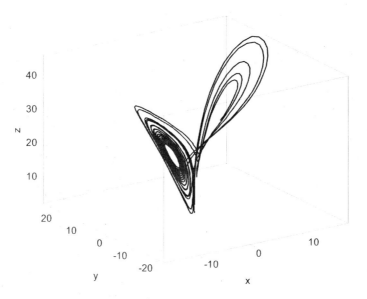

FIGURE 3.10
The chaotic attractor for the Lorenz system

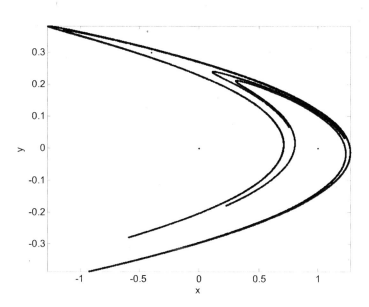

FIGURE 3.11
The Hénon attractor

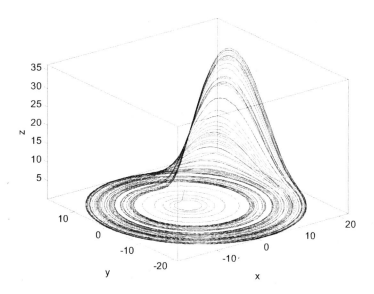

FIGURE 3.12
The Rössler attractor

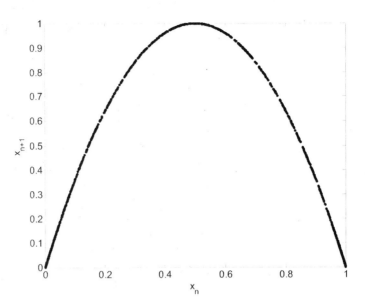

FIGURE 3.13
The lagged phase space for the logistic map using an embedding dimension of
two and a lag of one time instant

pattern in phase space that is not evident from the time series representation
in Fig. 3.1. In this notation x_{n+1} represents the signal x_n lagged by one time
instant. The embedding dimension is said to be two because the plot is in
two dimensions. Figure 3.14 shows a more detailed pattern in a lagged phase
space (embedding dimension of three and lag of one time instant). The graph
is a plot of x_{n+2} versus x_{n+1} versus x_n. In this notation x_{n+2} represents the
signal x_n lagged by two time instants (or the signal x_{n+1} lagged by one time
instant). The mathematical formalization will be discussed briefly in the next
section.

Phase space plots are usually unable to show the internal features of the
attractor. Moreover, often attractors have more than three dimensions, in
which case they do not have simple visual representations. Therefore, in addition to the phase space plots, related lower-dimensional representations such

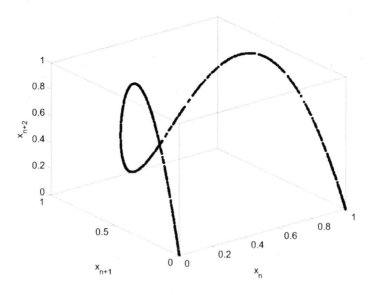

FIGURE 3.14
The lagged phase space for the logistic map using an embedding dimension of
three and a lag of one time instant

as Poincaré sections, return maps, next-amplitude plots, and difference plots
may be studied to obtain a more comprehensive understanding of the attractor. *Poincaré sections* are slices through the attractor which give a better
understanding of the trajectory at any particular cross-sectional plane. The
mathematical function that yields information on the next intersection of the
trajectory with the Poincaré section under consideration is called a *return
map*. The return map is based on the location of the previous intersection in
phase space. The *next-amplitude plots* are representative of the complete attractor as opposed to only a slice through it and show the lagged phase space
of oscillation peaks (or local maxima). Finally, *difference plots* show lagged
data for differences in successive observations. They are very similar to lagged
phase space plots but the axes of the plots reflect the differences in successive
observations. For example, a difference plot may be plotted with $x_{n+2} - x_{n+1}$
as the y-axis and $x_{n+1} - x_n$ as the x-axis.

3.3   Chaos Analysis

3.3.1   Measures of Chaos

The most important representation of a chaotic system is the governing rule or map that determines how the system evolves forward in time. Two main characteristics in the analysis of a system and therefore its attractor are complexity and chaoticity. Complexity is a measure of the geometric properties of the system and is characterized by the magnitude of the attractor dimension. *Dimension* can have different interpretations depending on the context. In addition to the regular Euclidean, topological, and vector definitions, a dimension can also be defined as the scaling exponent in a power law. As a scaling exponent, the value of a dimension need not be an integer value. For instance, an attractor embedded in a phase space of embedding dimension 2 would typically have a scaling exponent between 1 and 2. The most common types of scaling exponent dimensions in chaos theory are similarity, capacity, Hausdorff, information, correlation, and fractal dimensions (Bullmore et al., 1992; Cao, 1997; Williams, 1997; Jiang and Adeli, 2003; Notley and Elliott, 2003). The correlation dimension characterizes the attractor at a fine resolution and is computationally efficient and is therefore used as the measure of complexity in the applications presented in this book.

The chaoticity of the attractor is a measure of the convergence or divergence of nearby trajectories in phase space. In a chaotic system two points close together initially in phase space could have very different final outcomes, as shown in Fig. 3.2. This behavior is known as the *butterfly effect*. Therefore, a divergence in the trajectories could suggest the presence of a chaotic attractor. On the other hand, if the trajectories converge, it would imply a non-chaotic attractor. Measures such as the largest Lyapunov exponent,

Kolmogorov-Sinai entropy, and mutual information and redundancy coefficients are typically used to quantify chaos in a system. Of these, the largest Lyapunov exponent has been shown to be a reliable measure and therefore is used in the applications in this book (Iasemidis et al., 1995).

3.3.2 Preliminary Chaos Analysis - Lagged Phase Space

A signal can be represented as a time series vector composed of individual data points (for example, single voltage readings by an electrode at various time instants) as:

$$\mathbf{X} = \{x_1, x_2, \dots, x_N\} \tag{3.1}$$

where N is the total number of data points and the subscript indicates the time instant. Assuming a selected time lag m, $\mathbf{X}_T$ represents a time series vector that contains all data points in $\mathbf{X}$ from time instant T to $N - m$ (i.e., $x_T, x_{T+1}, \dots, x_{N-m}$) and $\mathbf{X}_{T+m}$ represents a time series vector that contains all data points in $\mathbf{X}$ from time instant $T + m$ to N (i.e., $x_{T+m}, x_{T+m+1}, \dots, x_N$), then the graph of $\mathbf{X}_{T+m}$ versus $\mathbf{X}_T$ is known as the pseudo or lagged phase space with lag m and embedding dimension 2. Both $\mathbf{X}_T$ and $\mathbf{X}_{T+m}$ are subsets of $\mathbf{X}$ and contain $N - T - m + 1$ data points each. Employing an embedding dimension of 3 would result in a lagged phase space with $\mathbf{X}_{T+2m}$, $\mathbf{X}_{T+m}$, and $\mathbf{X}_T$ as the three axes. The lagged phase space is important for identifying the temporal evolution of the signal. In order to create the lagged state space, two parameters must be identified: the optimum lag and the minimum embedding dimension.

In order to find the optimum lag, the signal can be tested for autocorrelation or mutual information which are measures of the extent of overlap between the information contained in $\mathbf{X}_T$ and $\mathbf{X}_{T+m}$. An effective lagged phase space requires the overlap to be minimal so as to avoid a direct relationship between $\mathbf{X}_T$ and $\mathbf{X}_{T+m}$. By definition if $\mathbf{X}_T$ and $\mathbf{X}_{T+m}$ contain completely overlapping information or are exactly identical such as at $m = 0$, then the lagged phase space would be a straight line at a 45° angle. The optimum lag m_0 is termed so because it has to be large enough for $\mathbf{X}_T$ and $\mathbf{X}_{T+m}$ to yield minimal overlapping information without being so large that the number of data points in the signals compared becomes too small. Since the total number of data points in the EEG signal is limited to N, for a given lag m, $\mathbf{X}_T$ and $\mathbf{X}_{T+m}$ each contain $N - m$ data points. If the value of m is too large, then the size of the vectors $\mathbf{X}_T$ and $\mathbf{X}_{T+m}$ becomes very small and some important features may be lost.

If the data have too much overlapping information, the cause of the overlap should be removed. For instance, in a periodic signal there may be extensive overlap between $\mathbf{X}_T$ and $\mathbf{X}_{T+m}$ even for large values of the lag m. In this case, the periodicity must be removed from the signal. This can be achieved with techniques such as standardization, filtering, and decomposition which may be based on statistical methods or on more advanced signal processing methods such as Fourier, wavelet, and fractal analyses.

The autocorrelation coefficient R_m for lag m is mathematically defined as:

$$R_m = \frac{\sum_{i=1}^{N-m} (\mathbf{X}_T(i) - \overline{\mathbf{X}})(\mathbf{X}_{T+m}(i) - \overline{\mathbf{X}})}{\sum_{i=1}^{N} (\mathbf{X}(i) - \overline{\mathbf{X}})^2} \qquad (3.2)$$

where N is the number of observations and $\overline{\mathbf{X}}$ is the mean of the observed values of the physical feature $\mathbf{X}$ defined over time as:

$$\overline{\mathbf{X}} = \frac{\sum_{i=1}^{N} \mathbf{X}(i)}{N} \qquad (3.3)$$

Varying the value of m yields autocorrelation coefficients (R_m) for different values of lag. These are graphically represented by plotting autocorrelation versus lag in a correlogram. The optimum lag is the first local minimum on the graph of R_m versus m.

The mutual information coefficient I_m for lag m is mathematically defined as (Williams, 1997):

$$I_m = \sum_{i=1}^{N_S} \sum_{j=1}^{N_S} P[\mathbf{X}_T(i), \mathbf{X}_{T+m}(j)] \log_2 \frac{P[\mathbf{X}_T(i), \mathbf{X}_{T+m}(j)]}{P[\mathbf{X}_T(i)] P[\mathbf{X}_{T+m}(j)]} \qquad (3.4)$$

where N_S is the number of probability states or bins, $P[\mathbf{X}_T(i)]$ is the probability of $\mathbf{X}_T$ belonging to the ith probability state, and $P[\mathbf{X}_{T+m}(j)]$ is the probability of $\mathbf{X}_{T+m}$ belonging to the jth probability state. The term $P[\mathbf{X}_T(i), \mathbf{X}_{T+m}(j)]$ is the joint probability of $\mathbf{X}_T$ belonging to the ith probability state and $\mathbf{X}_{T+m}$ belonging to the jth probability state simultaneously. As a standard practice in information theory, since the logarithm has a base of 2 the unit of the mutual information coefficient I_m is a binary digit or *bit*. The probability of $\mathbf{X}_T$ belonging to the ith probability state $P[\mathbf{X}_T(i)]$ is computed as (Williams, 1997):

$$P[\mathbf{X}_T(i)] = \frac{n(i)}{N - T - m + 1} \qquad (3.5)$$

where $n(i)$ is the number of data points in $\mathbf{X}_T$ belonging to the ith probability state and the denominator is the total number of data points in $\mathbf{X}_T$. Other probability values are computed similarly. Varying the value of m yields mutual information coefficients (I_m) for different values of the time lag. The optimum lag is the first local minimum on the graph of I_m versus m.

After the selection of the optimum lag, a common method for estimating the minimum embedding dimension for the phase space of the signal is Cao's method (Cao, 1997). Next, the ith time-delay vector is reconstructed from the

EEG signal $\mathbf{X}$ using the optimum lag value, m_0, and the estimated minimum embedding dimension, d_M. The reconstructed time delay vector $Y_i(d)$ in the lagged phase space has the following form:

$$\mathbf{Y}_i(d) = \{x_i, x_{i+m_0}, \ldots, x_{i+m_0(d-1)}\} \qquad (3.6)$$

where $i = 1, 2, \ldots, N - m_0(d-1)$, d is the embedding dimension, and x_i is the ith data point in the EEG signal. The underlying principle of this method is that if d is a *true* embedding dimension, then two points that are close to each other in the d dimensional phase space will remain close in the $d + 1$ dimensional phase space. Any two points satisfying the above condition are known as *true* neighbors (Cao, 1997).

The method is applied repeatedly starting with a low value of the embedding dimension d and then increasing it until the number of false neighbors decreases to zero, or equivalently, Cao's embedding function defined as (Cao, 1997):

$$E(d) = \frac{1}{N - m_0 d} \sum_{i=1}^{N - m_0 d} a_i(d) \qquad (3.7)$$

becomes constant. In Eq. (3.7)

$$a_i(d) = \frac{\left\| \mathbf{Y}_i(d+1) - \mathbf{Y}_{n(i,d)}(d+1) \right\|}{\left\| \mathbf{Y}_i(d) - \mathbf{Y}_{n(i,d)}(d) \right\|} \qquad (3.8)$$

where $i = 1, 2, \ldots, N - m_0 d$, and $\mathbf{Y}_{n(i,d)}(d)$ is the nearest neighbor of $\mathbf{Y}_i(d)$ in the d dimensional space. The proximity of two neighbors for deciding the nearest neighbor is based on a measure of distance computed using the maximum norm function denoted by $\|*\|$ in Eq. (3.8). The embedding function is modified to model the variation from d to $d + 1$ by defining another function, $E_1(d)$, which converges to 1 in the case of a finite dimensional attractor as follows (Cao, 1997):

$$E1(d) = \frac{E(d+1)}{E(d)} \tag{3.9}$$

The minimum embedding dimension, d_M, is identified from the graph of $E1(d)$ versus d as the value of d at which the value of $E1(d)$ approaches 1. However, in certain cases this may occur even with truly random signals. In order to distinguish deterministic data from truly random signals, another function is defined as (Cao, 1997):

$$E2(d) = \frac{E^*(d+1)}{E^*(d)} \tag{3.10}$$

where $E^*(d)$ is defined as:

$$E^*(d) = \frac{1}{N - m_0 d} \sum_{i=1}^{N-m_0 d} \left| x_{i+m_0 d} - x_{n(i,d)+m_0 d} \right| \tag{3.11}$$

in which $x_{n(i,d)+m_0 d}$ is the nearest neighbor of $x_{i+m_0 d}$. From an examination of the graph of $E2(d)$ versus d, a constant value of 1 for $E2(d)$ for different values of d indicates a truly random signal. The signal is found to be deterministic if the value of $E2(d)$ is not equal to 1 for at least one value of d. The embedding dimension for the phase space is set to the minimum embedding dimension. The lagged phase space is constructed based on the identified values of the optimum time lag and the embedding dimension.

3.3.3 Final Chaos Analysis

Correlation Dimension (CD) of the Attractor

Given a finite signal represented by $N_C = N - m_0 d_M$ points denoted by $\mathbf{Y}_1(d_M), \ldots, \mathbf{Y}_{N_C}(d_M)$ in phase space, the correlation sum, C_ϵ, for a measuring circle of radius ϵ, is mathematically defined as (Williams, 1997):

$$C_\epsilon = \lim_{N_C \to \infty} \frac{1}{N_C^2} \sum_{i=1}^{N_C} \sum_{j=1}^{N_C} G\left(\epsilon - |\mathbf{Y}_i(d_M) - \mathbf{Y}_j(d_M)|\right) \tag{3.12}$$

where $\mathbf{Y}_i(d_M)$ and $\mathbf{Y}_j(d_M)$ are the lagged phase space locations of the ith and jth points respectively for the selected embedding dimension d_M. In general, ϵ is the radius of the measuring unit. For an embedding dimension of 2, the measuring unit is a circle whereas for an embedding dimension of 3, the measuring unit is a sphere.

The function G is the Heaviside function which returns a positive count only when the jth point lies within a distance of ϵ from the ith point, i.e.:

$$\epsilon - |\mathbf{Y}_i(d_M) - \mathbf{Y}_j(d_M)| > 0 \qquad (3.13)$$

The correlation dimension (ν) is approximated from the slope of the plot of the log of the correlation sum (C_ϵ) versus the measuring radius (ϵ). The mathematical relation is presented as:

$$C_\epsilon \propto \epsilon^\nu \qquad (3.14)$$

Given any finite signal represented by N_C points in the phase space, there can be a total of $N_C(N_C - 1)/2$ pairwise distances (represented mathematically as $\|x_i - x_j\|$ where $i \neq j$). In the wavelet-chaos algorithm employed for the applications in this book, the CD is computed directly using the Takens estimator as (Cao, 1997; Borovkova et al., 1999):

$$\nu = -\left[\frac{2}{N_C(N_C - 1)} \sum_{i=1}^{N_C} \sum_{j=1}^{N_C} \log \left(\frac{|\mathbf{Y}_i(d_M) - \mathbf{Y}_j(d_M)|}{\epsilon} \right) \right] \qquad (3.15)$$

Largest Lyapunov Exponent (LLE)

As discussed previously in Section 3.3.1, an important characteristic of a chaotic system is trajectory divergence. Trajectory divergence is defined as the change in the distance between two neighboring points in the lagged phase space after a given time. Lyapunov exponents are measures of the *rate* of trajectory divergence in a system. As the system evolves from time zero to infinity, local Lyapunov exponents are computed continuously as (Williams, 1997):

$$\lambda_i = \log_e |f'(\mathbf{Y}_i(d_M))| \tag{3.16}$$

where $f'(\mathbf{Y}_i(d_M))$ is the rate of divergence of two neighboring trajectories at point $\mathbf{Y}_i(d_M)$ in the phase space. The standard Lyapunov exponent (λ) is computed as the mathematical average of the local Lyapunov exponents along each dimension of the attractor as:

$$\lambda = \lim_{n \to \infty} \frac{1}{n} \sum_{i=0}^{n-1} \lambda_i \tag{3.17}$$

where n is the number of time steps of the evolving system.

The number of standard Lyapunov exponents is equal to the embedding dimension of the attractor. For the system to be chaotic, the trajectories must diverge along at least one dimension of the attractor which implies that at least one of the standard Lyapunov exponents must be positive. As a result, by definition, the LLE ($\lambda_{\max}$) must be greater than zero in a chaotic system.

A more direct method for computing the LLE is Wolf's method (Wolf et al., 1985; Rosenstein et al., 1993; Hilborn, 2001). According to Wolf's method, the average trajectory divergence, D_T, of the attractor after a given time T (known as prediction length) is expressed mathematically as:

$$D_T = \frac{1}{N_S} \sum_{i=1}^{N_S} \left| \frac{\mathbf{Y}_{i+T}(d_M) - \mathbf{Y}'_{i+T}(d_M)}{\mathbf{Y}_i(d_M) - \mathbf{Y}_i'(d_M)} \right| \tag{3.18}$$

where $\mathbf{Y}_i(d_M)$ and $\mathbf{Y}_i'(d_M)$ are the neighboring points on separate trajectories in the phase space, and $\mathbf{Y}_{i+T}(d_M)$ is the location of the point that evolved from $\mathbf{Y}_i(d_M)$ along the trajectory. The prediction length, T, is measured in increments of time used for the signal.

The LLE (λ_{max}) is subsequently computed as the slope of the graph of the natural logarithm of trajectory divergence, D_T, versus the prediction length, T. This relationship is expressed mathematically as:

$$D_T = D_0 e^{T\lambda_{\mathrm{max}}} \qquad (3.19)$$

where D_0 is the initial divergence. In the applications in this book, a modification of Wolf's method reported in Iasemidis et al. (2000a) is used in which the parameters are adaptively estimated to better account for the non-stationary nature of real-world signals.

4

Classifier Designs

4.1 Data Classification

The last two chapters focused on techniques for denoising signals and extracting or highlighting meaningful information from the signals including higher dimensional patterns that may not be evident from a visual inspection. In general, the *meaningfulness* of the information is subjective and depends on the problem being solved. Ultimately, the success of any strategy depends on the achievement of the objectives. Therefore, the next step is to examine this information in the context of these objectives. The strategy adopted in the approaches presented in this book involves the classification or organization of the data into meaningful groups.

Each meaningful *feature* extracted from the signal represents an aspect of the signal and is used (alone or in conjunction with other such features) to classify the signal. For this reason, such a feature is also called a classification parameter. A set of such features constitutes the *feature space*. The number of dimensions of the feature space is equal to the number of features. In other words, the original signal space is transformed into the feature space, which may be more suitable for signal classification. Assuming there are P features, each signal is represented in this feature space by one P-dimensional data point.

The classification process can be unsupervised or supervised. Unsupervised classification involves classifying the data points into groups solely on the basis of the similarity (or closeness) of the data points to each other. No information is available *a priori* regarding the groups themselves or the assignment of data points to the groups. Supervised classification, on the other hand, involves training the classifier to recognize that certain data points belong to certain groups. The data is divided into two sets - training data and testing data. The training data is presented to the classifier along with the known group assignments. Generally, based on the presented information, the classifier algorithm selects parameters and rules that appropriately model the classification of the training data. The same rules and parameters are then applied to the testing data in order to classify them into the appropriate groups and evaluate the accuracy of the classification method.

4.2   Cluster Analysis

Cluster analysis is an unsupervised learning algorithm. An important issue in cluster analysis is defining the similarity or *proximity* between each pair of points by a measure of distance such as the Euclidean distance, city block metric, or the Mahalanobis distance (Dillon and Goldstein, 1984; Kachigan, 1984). The Mahalanobis distance is used to better account for any correlation in the data.

If $\mathbf{F}$ represents the $N \times P$ matrix of data points in the P-dimensional feature space, where N is the total number of data points, then the mean-corrected data matrix (to center the data around the mean) may be expressed as:

$$\mathbf{F}_d = \mathbf{F} - \overline{\mathbf{F}} \tag{4.1}$$

where $\overline{\mathbf{F}}$ is the mean vector of size $1 \times P$. The Euclidean distance $d(i,j)$ between the ith and the jth points represented by $\mathbf{F}(i)$ and $\bar{\mathbf{F}}(j)$ is defined as:

$$\mathbf{d}(i,j) = [\mathbf{F}(i) - \mathbf{F}(j)][\mathbf{F}(i) - \mathbf{F}(j)]^T \tag{4.2}$$

where the superscript T denotes the transpose of the matrix. The Mahalanobis distance $\mathbf{d}(i,j)$ between the two points is defined as (Dillon and Goldstein, 1984):

$$\mathbf{d}(i,j) = [\mathbf{F}(i) - \mathbf{F}(j)]\mathbf{C}^{-1}[\mathbf{F}(i) - \mathbf{F}(j)]^T \tag{4.3}$$

where $\mathbf{C}$ is the $P \times P$ sample covariance matrix expressed as:

$$\mathbf{C} = \frac{\mathbf{F}_d{}^T \mathbf{F}_d}{N - 1} \tag{4.4}$$

The $N \times N$ pair-wise distance matrix $\mathbf{d}$ (whether Euclidean or Mahalanobis) is reduced to the similarity matrix $\mathbf{Y}$ which is a $1 \times \mathrm{C}(N,2)$ row vector containing the distances between each pair of objects, where $\mathrm{C}(N,2)$ represents the number of distinct pairs that may be formed out of N objects.

In order to cluster the objects, the hierarchical single-linkage or nearest-neighbor method is applied. This process is shown in Fig. 4.1. The two objects closest to each other with respect to their proximity metric are paired together in one binary cluster. If any of the remaining unclustered objects has lesser distance to this cluster (characterized by the distance to the nearest object in the cluster) than to the other unclustered objects, it forms a binary cluster with the cluster in the previous step. Otherwise, it is paired with the closest unclustered object into a separate binary cluster. If two clusters are closer to each other than to any other unclustered object, then the two clusters are grouped together in a bigger cluster. This process continues until all the objects are clustered in one big cluster that contains all the smaller clusters. This binary cluster tree is represented by a dendrogram, which hierarchically

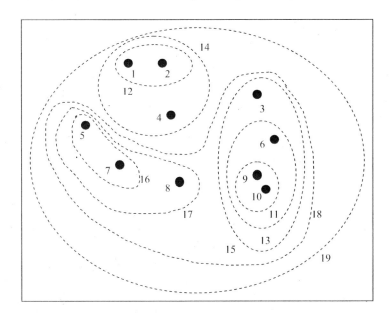

FIGURE 4.1
Example of the binary clustering process

summarizes the clustering stages, as shown in Fig. 4.2. This results in an $(N-1) \times 3$ linkage matrix P where each row represents a link or a binary cluster and contains the link characteristics: start node, end node, and link length. The binary clusters formed at each stage are considered as nodes for the next clustering stage and are numbered as $N+1, N+2, \ldots, 2N+1$.

Clusters at a particular level are said to be dissimilar if the length of a link between clusters or objects at that level in the dendrogram differs from the length of the links below it. The inconsistency coefficient of a link measures this dissimilarity by comparing the link length with the average of the link lengths at the same stage of clustering (which includes the link itself and the links l levels below it in the hierarchy where l is known as the depth of the comparison). For a link $q - N$, where $q \in N+1, N+2, \ldots, 2N+1$, the inconsistency coefficient is mathematically defined as (Ghosh-Dastidar and

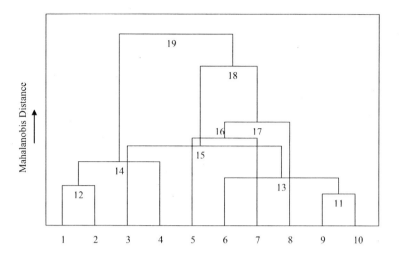

FIGURE 4.2
Dendrogram for the binary clustering example shown in Fig. 4.1

Adeli, 2003):

$$\mathbf{I_C}(q - N) = \frac{\mathbf{P}(q, 3) - \overline{\mathbf{P}_l(q)}}{\sigma_{l, \mathbf{P}(q)}} \tag{4.5}$$

where for a depth of $l = 2$ levels,

$$\overline{\mathbf{P}_2(q)} = \frac{\mathbf{P}(q, 3) + \mathbf{P}(\mathbf{P}(q, 1) - N, 3) + \mathbf{P}(\mathbf{P}(q, 2) - N, 3)}{r} \tag{4.6}$$

in which r is the number of clusters excluding the original nodes and:

$$\sigma_{l, \mathbf{P}(q)} = \left[\frac{[\mathbf{P}(q, 3) - \overline{\mathbf{P}_2(q)}] + [\mathbf{P}(\mathbf{P}(q, 1) - N, 3) - \overline{\mathbf{P}_2(q)}]}{r - 1} \right.$$
$$\left. + \frac{[\mathbf{P}(\mathbf{P}(q, 2) - N, 3) - \overline{\mathbf{P}_2(q)}]}{r - 1} \right]^{1/2} \tag{4.7}$$

This results in an $(N - 1) \times 1$ inconsistency coefficient matrix representing the similarities or dissimilarities in the clusters formed at a given level. A high inconsistency coefficient often implies class boundaries. Finally, the data set is divided predominantly into two clusters having an inconsistency coefficient below the selected threshold level. Besides the major clusters, the clustering process may highlight the existence of outliers which are usually grouped separately.

4.3   *k*-Means Clustering

k-means clustering is a variation of the unsupervised clustering method. k-means clustering is also an unsupervised clustering method but differs from regular clustering in that the number of clusters is fixed at k. In that sense, it is not completely unsupervised because *a priori* information is available regarding the number of clusters. Since all the data points are to be divided into k clusters, the process is initiated by an arbitrary selection of k points. These k points form the seeds of the k clusters in the analysis.

At this initial stage, the cluster centroids are the points themselves (since the cluster only contains one point). The next point closest in terms of the selected proximity metric (termed *nearest neighbor*) to any one of these clusters is paired with that cluster in one binary cluster. This cluster now contains two points and a new centroid is computed. If any one of the remaining unclustered points has a shorter distance to this cluster (defined by its proximity to the centroid of the cluster) than to the other clusters, it forms a binary cluster with the same cluster. Otherwise, it is paired with the nearest cluster. Addition of a point to a cluster changes the centroid of the cluster which may lead to changes in cluster assignments. This process is repeated until the cluster centroids stop changing and all points are divided into the predetermined k clusters.

4.4 Discriminant Analysis

The aim of discriminant analysis is to increase the separation between the different groups of data points by mapping or projecting the input feature space to a lower-dimensional output space where the intra-group variance is minimized and the inter-group variance is maximized. Discriminant analysis is a supervised classification process and requires training data (where the group assignments for the data points are known). Assuming an underlying normal distribution for the data, if the mean of the entire $N_{\text{TR}} \times P$ training set input, $\mathbf{F_R}$ (consisting of N_{TR} data points in the training set and P features), is represented by the vector $\mu_{\mathbf{R}}$ and the group mean of the input data points belonging to the jth group, $\mathbf{F}_{\mathbf{R}_j}$ (consisting of N_j data points), is represented by the vector $\mu_{\mathbf{R}_j}$, then the $P \times P$ intra-group variance matrix is computed as (Fukunaga, 1990):

$$\mathbf{S_W} = \sum_{j=1}^{3} \frac{N_j}{N_{\text{TR}}} [\mathbf{F}_{\mathbf{R}_j} - \mu_{\mathbf{R}_j}][\mathbf{F}_{\mathbf{R}_j} - \mu_{\mathbf{R}_j}]^T \tag{4.8}$$

where the superscript T denotes the transpose of the matrix. The $P \times P$ inter-group variance matrix is computed as:

$$\mathbf{S_B} = \sum_{j=1}^{3} \frac{N_j}{N_{\text{TR}}} [\mu_{\mathbf{R}_j} - \mu_{\mathbf{R}}]^T [\mu_{\mathbf{R}_j} - \mu_{\mathbf{R}}] \tag{4.9}$$

The maximization-minimization objective is achieved by maximizing the sum of the eigenvalues of $\mathbf{S_W}^{-1}\mathbf{S_B}$ (Fukunaga, 1990). This forms the basis of linear discriminant analysis (LDA) where the class boundaries are hyperplanes. This is the most basic form of LDA and is referred to as ELDA (E stands for Euclidean) in this book. The Euclidean distance may not yield accurate results when parameters are statistically correlated. In such cases, to account for correlation between the parameters constituting the feature space, the Mahalanobis distance is used instead. This form of LDA is referred to as MLDA in this book.

LDA is based on the assumption that the covariance matrices for different groups are similar. Quadratic discriminant analysis (QDA) is employed when the covariance matrices are considerably different. The QDA classification function accounts for unequal covariance matrices, resulting in class boundaries that are second order surfaces. The disadvantages of QDA include an increased computational burden, the need for a larger training dataset, and a greater sensitivity to deviations from normality and incorrect classifications in the training set.

4.5   Principal Component Analysis

Principal component analysis (PCA) is another statistical method used primarily to transform the input space, usually into a lower dimensional space. In essence, the coordinate system is rotated using a linear transformation. The axes (or components) of the new coordinate system are the eigenvectors that describe the data set and therefore are linear combinations of the original axes. The primary axis or principal component is selected to represent the direction of maximum variation in the data. The secondary axis, orthogonal to the primary axis, represents the direction of the next largest variation in the data and so on. In the reoriented space, most of the variation in the data is concentrated in the first few components. Consequently, the components that account for most of the variability are retained whereas the remaining components are ignored. As a result, dimensionality can be reduced without compromising the accuracy of data representation in any significant way (Lee and Choi, 2003).

Unlike discriminant analysis, which can be used as a classifier or as a data preprocessing step prior to classification, PCA is used primarily for data preprocessing such as dimensionality reduction and noise elimination. Generally, assuming that the training data is represented by the matrix $\mathbf{F_R}$, the following steps are involved:

1. The centroid of the data points, denoted by the $1 \times P$ vector $\mu_{\mathbf{R}}$, is obtained, where P is the number of features.

2. The origin of the coordinate system is moved to the centroid of the data points by subtracting $\mu_{\mathbf{R}}$ from each point in $\mathbf{F_R}$.

3. The pairwise covariance between all parameters constitutes the columns of the shifted training input is computed. This is represented by a $P \times P$ covariance matrix, denoted by $cov(\mathbf{F_R} - \mu_{\mathbf{R}})$. The covariance matrix is symmetric and the diagonal elements of the matrix are the variance values for the P features.

4. The eigenvectors and eigenvalues of the covariance matrix are computed, resulting in P eigenvectors of size $P \times 1$ and the P corresponding eigenvalues. The eigenvectors are mutually orthogonal and represent the axes of the new coordinate system. The eigenvector corresponding to the maximum eigenvalue is the principal component.

5. The importance of the eigenvector as a representation of the data variability decreases with a decrease in the corresponding eigenvalue. Therefore, eigenvectors corresponding to the lowest eigenvalues are discarded. The remaining P_{SEL} eigenvectors are arranged column-wise in the order of decreasing eigenvalues to form the $P \times P_{\mathrm{SEL}}$ eigenvector matrix, $\mathbf{E}$. How many eigenvectors to keep is found by trial and error.

4.6   Artificial Neural Networks

Artificial neural networks (ANNs) are computational models of learning that are inspired by the biology of the human brain. ANNs consist of neurons (also called nodes or processing elements) which are interconnected via synapses (also called links). From a functional perspective, ANNs mimic the learning abilities of the brain and can, ideally, be trained to recognize any given set of inputs by adjusting the synaptic weights. A properly trained network, in principle, should be able to apply this learning and respond appropriately to completely new inputs. ANNs, however, are based on highly simplified brain dynamics, which makes them much less powerful than their biological counterpart. Nevertheless, ANNs have been used as powerful computational tools to solve complex pattern recognition, function estimation, and classification problems not amenable to other analytical tools (Adeli and Hung, 1995; Adeli and Park, 1998; Adeli and Karim, 2000; Adeli, 2001; Ghosh-Dastidar and Adeli, 2003; Adeli and Karim, 2005; Adeli et al., 2005a,b; Adeli and Jiang, 2006; Ghosh-Dastidar and Adeli, 2006).

The most common application of ANNs is supervised classification and therefore requires separate training and testing data. Since the learning is usually performed with the training data, the mathematical formalization is based on the training data. Two of the most commonly used neural network architectures are discussed in this chapter. The selected architectures are both feedforward architectures where information is transmitted across neurons starting from the input layer across the network to the output layer. These architectures form the basis of advanced architectures and models that will be discussed later in this book. The input training matrix for the classifier is denoted by $\mathbf{F_R}$ and the desired output vector by $\mathbf{O_R}$. The least mean squares method is used to minimize the error function.

4.6.1 Feedforward Neural Network and Error Backpropagation

The layers of a feedforward ANN classifier are numbered from right to left starting with $l = 0$ for the output layer. There can be any number of hidden layers. Architectures with one and two fully connected hidden layers are shown in Figs. 4.3 and 4.4, respectively. The total input to the jth node of any layer l representing the layer that is l layers before the final output layer is mathematically written as (Bose and Liang, 1996):

$$i_j^l = \sum_i y_i^{l+1} w_{ij}^l - \theta_j^l \qquad (4.10)$$

where y_i^{l+1} is the output of the ith node in layer $l+1$, w_{ij}^l is the weight of the connection from the ith node in layer $l+1$ to the jth node in layer l and θ_j^l is the node bias of the jth node in layer l. A simple input integration is shown in Fig. 4.5 where $\mathbf{i}$ represents the input and $\mathbf{w}$ represents the weights. For the sake of simplicity in this figure, only the subscript representing the input neuron number is retained in the notation even though the input neurons are not shown. The summation symbol $\sum$ denotes that the weighted inputs are summed before transformation using the activation function f.

The input to each node in the hidden or output layer l is the same as the output from layer $l+1$ represented in vector notation by the $N_{l+1} \times 1$ vector $\mathbf{Y}_{l+1}$ where N_{l+1} is the number of nodes in layer $l+1$. The $N_l \times 1$ weighted input vector to the hidden layer l is expressed as:

$$\mathbf{I}_l = \mathbf{W}_l^T \mathbf{Y}_{l+1} \qquad (4.11)$$

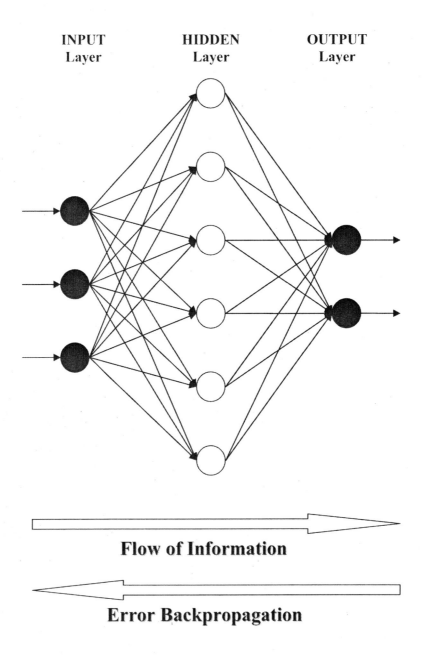

FIGURE 4.3
A fully connected feedforward ANN with one hidden layer

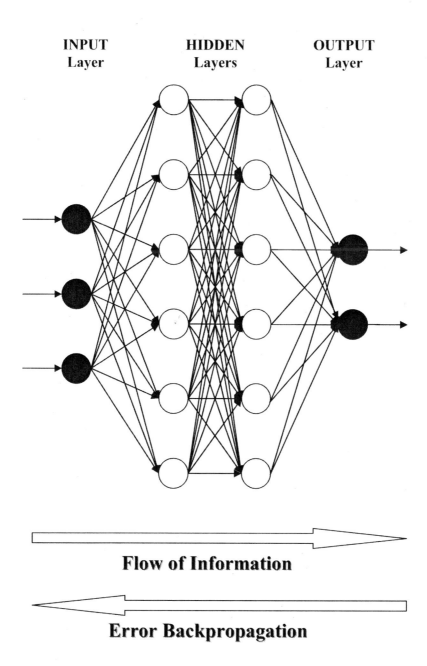

FIGURE 4.4
A fully connected feedforward ANN with two hidden layers

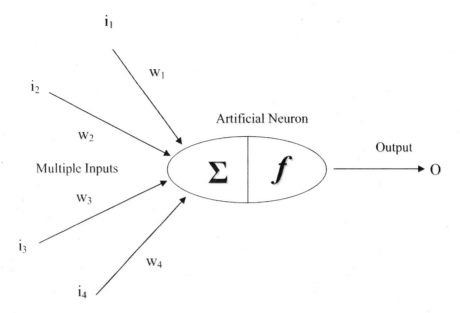

FIGURE 4.5
Simple input integration by an ANN neuron

where $\mathbf{W}_l$ is the $N_{l+1} \times N_l$ weight matrix for layer l. If any *bias* is associated with any of the nodes (to enable such nodes to fire preferentially or non-preferentially compared to others), this equation is modified as:

$$\mathbf{I}_l = \mathbf{W}_l{}^T \mathbf{Y}_{l+1} - \boldsymbol{\theta}_l \tag{4.12}$$

where $\boldsymbol{\theta}_l$ is the $N_l \times 1$ bias vector for the nodes in layer l. Since the input to the input layer is $\mathbf{F_R}$, the input to the hidden layer immediately succeeding the input layer is computed as:

$$\mathbf{I}_l = \mathbf{W}_l{}^T \mathbf{F_R} - \boldsymbol{\theta}_l \tag{4.13}$$

The output of the jth node in layer l is computed using the log-sigmoid activation function as (Adeli and Samant, 2000):

$$\mathbf{Y}_l(j) = \frac{1}{1 + e^{-\mathbf{I}_l(j)}} \tag{4.14}$$

Another common activation function used in feedforward ANNs is the tan-sigmoid function expressed as (Bose and Liang, 1996):

$$\mathbf{Y}_l(j) = \frac{2}{1 + e^{-\mathbf{I}_l(j)}} - 1 \tag{4.15}$$

The sum squared error in network output is computed as:

$$\mathbf{E} = \frac{1}{2}[\mathbf{Y_0} - \mathbf{O_R}]^T[\mathbf{Y_0} - \mathbf{O_R}] \tag{4.16}$$

Neural network learning is an iterative process where the weights are adjusted in every iteration to minimize the value of the error function. Typically, neural network learning is incremental where each iteration corresponds to one training instance. Batch processing is also possible where the weights are adjusted once all the training instances have been presented. In that case, the error functions and the error minimization algorithm must be adjusted accordingly. However, the concept remains essentially the same and therefore only incremental training will be described in this section. At this point it becomes necessary to denote the iteration number or the training instance with the index k. As a result, the error function for the kth training instance becomes:

$$\mathbf{E}(k) = \frac{1}{2}[\mathbf{Y_0}(k) - \mathbf{O_R}]^T[\mathbf{Y_0}(k) - \mathbf{O_R}] \tag{4.17}$$

This error is propagated backwards throughout the nodes of each layer and the required changes are made to the weights for each connection. The network keeps updating the weights and biases using the gradient-descent method until the error convergence condition is met. Checking for convergence implies checking for the minimum value of the error or performance function by computing its negative gradient or the direction of steepest descent. The computation of the gradient and updating of the weights occurs after each training instance according to the generalized delta update rule, which is expressed as (Bose and Liang, 1996):

$$\delta(k) = \mathbf{W}(k+1) - \mathbf{W}(k)$$
$$= -\eta \mathbf{G}(k) \tag{4.18}$$

where $\mathbf{W}(k+1)$ denotes the new weight vector and $\mathbf{W}(k)$ denotes the current weight vector for the network, η is the learning rate, and $\mathbf{G}(k)$ is the current gradient of the error function. The gradient is defined as (Bose and Liang, 1996):

$$\mathbf{G}(k) = \nabla \mathbf{E}(k)$$
$$= \frac{\partial \mathbf{E}(k)}{\partial \mathbf{W}} \tag{4.19}$$

The simple BP algorithm suffers from a very slow rate of convergence (Adeli and Hung, 1995). For faster convergence of the network, quasi-Newton algorithms replace the fixed learning rate η with an adaptive learning rate based on the current weights and biases. This modifies Eq. (4.18) to:

$$\delta(k) = \mathbf{W}(k+1) - \mathbf{W}(k)$$
$$= -\mathbf{H}^{-1}(k)\mathbf{G}(k) \tag{4.20}$$

where $\mathbf{H}^{-1}(k)$ is the inverse of the Hessian matrix of the error function for the kth training instance. The Hessian matrix is expressed as:

$$\mathbf{H}(k) = \nabla^2 \mathbf{E}(k)$$
$$= \frac{\partial^2 \mathbf{E}(k)}{\partial \mathbf{W}^2} \tag{4.21}$$

To reduce the large computational power required for calculating the Hessian matrix for feedforward networks, the Levenberg-Marquardt algorithm uses a numerical approximation of the Hessian matrix. Since the error function is a sum of squares, the Hessian matrix is approximated as:

$$\mathbf{H}(k) = \nabla \mathbf{E}(k)^T \nabla \mathbf{E}(k) \tag{4.22}$$

and the gradient in Eq. (4.19) is modified to:

$$\mathbf{G}(k) = \nabla \mathbf{E}(k)^T \mathbf{E}(k) \tag{4.23}$$

Substituting Eqs. (4.22) and (4.23), the update rule of Eq. (4.20) is changed to:

$$\delta(k) = \mathbf{W}(k+1) - \mathbf{W}(k)$$
$$= - \left[\nabla \mathbf{E}(k)^T \nabla \mathbf{E}(k) + \mu_n \mathbf{I} \right]^{-1} \nabla \mathbf{E}(k)^T \mathbf{E}(k) \tag{4.24}$$

where the term $\mu_n \mathbf{I}$ is added to ensure that the Hessian matrix is invertible. The value of the parameter μ_n is initially selected as a random number between 0 and 0.1 and subsequently decreases in value whenever the value of the error function decreases and vice versa (Hagan et al., 1996; Ghosh-Dastidar and Adeli, 2003). This attempts to ensure that the error function avoids entrapment in local mimima and reaches the global minimum. The training of the algorithm is stopped when the change in system error decreases to a preset value signifying convergence or when the value of μ_n exceeds a preset maximum. The weights of the connections and biases of the nodes at convergence are said to be the weights and biases of the trained neural network. Once the ANN is trained, the ANN is tested using the testing data $\mathbf{F_T}$.

4.6.2   Radial Basis Function Neural Network

The underlying feedforward methodology of the radial basis function neural network (RBFNN) is similar to that of the backpropagation neural network but there are three main differences. First, RBFNN has only one hidden layer which can have a maximum of N_{TR} nodes (equal to the number of training instances). Second, the output from the output layer is usually computed using a linear activation function instead of the sigmoid activation functions. Third, the method of computation of the weighted inputs and outputs for the hidden layer is different. The weighted input to the hidden layer is computed as the vector of Euclidean distances between the $P \times 1$ input vector for the kth training instance, $\mathbf{F_R}(k)$ (a row vector), and the $N_{l+1} \times N_l$ weight matrix $\mathbf{W}_l$ of links connecting the N_{l+1} input nodes to the N_l nodes in the hidden layer. Since there is only one hidden layer, N_{l+1} is equal to P and N_l is equal to N_{TR}. Therefore, the weighted input to the jth hidden node for the kth training instance is expressed as:

$$\mathbf{I}_l(j) = [\mathbf{W}_l(j) - \mathbf{F_R}(k)]^T [\mathbf{W}_l(j) - \mathbf{F_R}(k)] \qquad (4.25)$$

The activation function for the hidden layer is a Gaussian function in the following form (Bose and Liang, 1996; Ghosh-Dastidar and Adeli, 2003):

$$\mathbf{Y}_l(j) = e^{\mathbf{I}_l(j)\,\log_e(0.5)/p^2} \qquad (4.26)$$

where p is the spread of the RBF which affects the shape of the Gaussian function. If the weight vector of the jth hidden node, $\mathbf{W}_l(j)$, is equal to the input vector for the kth training instance, $\mathbf{F_R}(k)$, the weighted input is 0 which results in an output of 1. The factor $\log_e(0.5)$ in the exponent term is used to scale the output to 0.5 (the average of the limits of 0 and 1) when the weighted input is equal to the spread, p. The activation function can be formulated by any function that has a value of one in the center and values of zero (or asymptotically tending to zero) at the periphery. This includes a triangular basis even though a triangular function is not differentiable at the center.

RBFNN training also employs the least mean square error method. However, unlike the backpropagation neural network, each iteration involves the addition of a hidden layer node, j. The input weight vector for this node, $\mathbf{W}_l(j)$, is selected to be equal to the input vector of the training instance k, $\mathbf{F_R}(k)$, that produces the minimum mean square error. In essence, each hidden layer node is trained to recognize a specific input. For this reason, training accuracy of the RBFNN is always 100% and the maximum number of hidden layer nodes is equal to the number of instances of training data. However, this is usually not an optimal solution because it could result in a large number of redundant hidden layer nodes (depending on the data) and a large computational burden. Therefore, various optimization strategies are usually incorporated to limit the number of hidden layer nodes, as discussed further in Section 9.3 of this book.

Part II

Automated EEG-Based Diagnosis of Epilepsy

5

Electroencephalograms and Epilepsy

5.1 Spatio-Temporal Activity in the Human Brain

Human brain activity displays a wide range of activation patterns during both
normal and abnormal states. Normal states include physical states (such as
sleep, wakefulness, and exertion) as well as mental states (such as calmness,
happiness, and anger). Abnormal states, primarily observed in neurological
disorders and drug-induced imbalances, include seizures (in epilepsy) and dementia (seen in Alzheimer's disease and Lewy body disease). The list of possi-
bilities is endless. This variety is further compounded by three factors. First,
each state has varying degrees of magnitude which results in varying degrees
and, in some cases, regions of brain activation. Second, brain activity in any
state is modulated by high-level brain functions such as attention and cognitive processing. Third, at any instant of time, the overall brain activity is not
due to any one mental state but rather to a superimposition of a number of
different states.

The brain processes underlying these states are very complex and require
a coordinated and efficient interaction between multiple areas of the brain. To
facilitate the efficient processing of such a vast number of states and their combinations, the brain is functionally organized such that different states yield
three primary types of activation patterns. Spatial patterns involve activation

of different areas of the brain, whereas temporal patterns involve activation of identical areas of the brain but at different times or in different sequences. The third type is a combination of the above mentioned two types and therefore termed *spatio-temporal* patterns. Also, states that are functionally similar often show similar spatio-temporal patterns but the magnitude of activation of different areas of the brain is different.

Although there are a large number of activation patterns, there is significant overlap in these patterns as a result of this *efficient* functional organization. Therefore, it is very difficult to use these patterns to conclusively identify the state. In fact, a complete spatio-temporal investigation of brain activity using modalities such as functional imaging or electroencephalography (EEG) is required before even attempting such an identification. A fast developing research area is that of EEG-based non-invasive and practical diagnostic tools for the investigation of various brain states, especially abnormal ones observed in neurological disorders. The popularity of EEG-based techniques over imaging techniques stems from two main advantages: 1) the relative inexpensiveness of equipment compared with imaging techniques and 2) the convenience for patients in clinical applications.

5.2   EEG: A Spatio-Temporal Data Mine

The EEG is a representation of the electrical activity generated during neuronal firing (synchronous electrical discharges) by neurons in the cerebral cortex. This electrical activity is recorded as electric potentials using special electrodes which are either placed on the scalp directly above various key areas of the brain (scalp EEG) or surgically implanted on the surface of the brain (intra-cranial EEG). The electrodes record the *field potential*, i.e., the sum total of the electric potentials in the receptive (or recording) field of the electrode. Scalp EEG electrodes are small metal discs with good mechanical adhesion and electrical contact. In order to provide an accurate detection of the field potentials, the electrodes are designed to have low impedance (less than 5kΩ). In addition, a conductive gel is often used for *impedance matching* between the electrode and the human skin to improve the signal-to-noise ratio.

Since the first human EEG recordings by Hans Berger were published in 1929 (Niedermeyer, 1999), the device for recording and displaying EEG signals has evolved from simple galvanometers and paper strips to digital computers. Nowadays, the signals from the electrodes are sensed and even amplified to some extent prior to transmission to a digital computer. After the computer receives the signals, software solutions are used to further reduce noise and artifacts. After the post-processing is complete, one continuous graphical plot is generated for each electrode to represent the change in the field potential signal over time. A multi-channel EEG is a collection of such field potential signals from every electrode and represents the spatial distribution of the field potentials in the brain (Fisch, 1999). This makes the EEG an invaluable tool for characterization of the spatio-temporal dynamics of neuronal activity in the brain and therefore for the detection and diagnosis of neurological disorders. A clinically recorded EEG from a normal adult is shown in Fig. 5.1.

Both intra-cranial and scalp EEGs have advantages and disadvantages. Intra-cranial EEG has a high signal-to-noise ratio because the electrode is implanted inside the brain and is therefore less susceptible to artifacts and electromagnetic interference. Moreover, it records from a small receptive field within the target site which allows 1) better exclusion of signals from outside the target zone and 2) a higher spatial resolution. However, due to its invasive nature, intra-cranial EEG is used primarily for research studies. Scalp EEG is more common clinically because it is non-invasive. The most common configu-

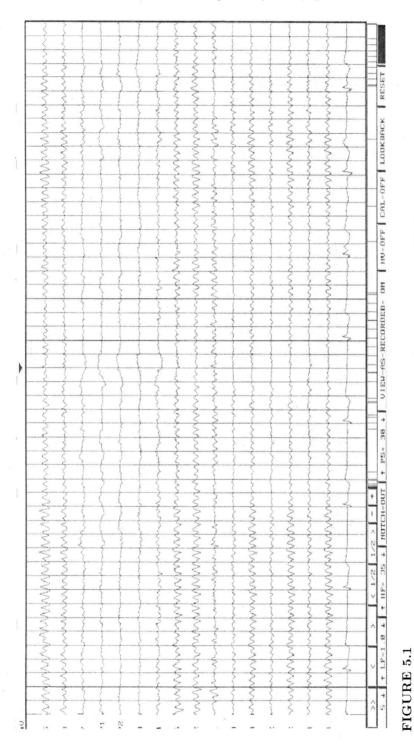

FIGURE 5.1
Normal adult EEG record for typical clinical diagnosis (courtesy Nahid Dadmehr, M.D.)

ration for electrode placement is the standard 10-20 configuration (American Clinical Neurophysiology Society, 2006). The electrode configuration and the clinically defined electrode locations are provided in Fig. 5.2. The configuration is so named to specify that the electrode spacing in the front-back direction is 10% of the total front-back distance along the skull and in the left-right direction is 20% of the total left-right distance along the skull. The modified and expanded 10-10 configuration (the naming may be interpreted in a similar manner) as specified by the American Clinical Neurophysiology Society (ACNS) guidelines is shown in Fig. 5.3. Sometimes, variations of these configurations are also employed, depending on preference and the nature of the disorder under investigation.

An EEG contains a wide range of frequency components. However, the typical frequencies of clinical and physiological interest lie in the range 0-30 Hz. Within this range, a number of approximate clinically relevant frequency bands or rhythms have been identified as follows (Kellaway, 1990):

- *Delta* (0-4 Hz): Delta rhythms are slow brain activities typically preponderant only in deep sleep stages of normal adults. Otherwise, they may be indicative of pathologies.

- *Theta* (4-7 Hz): Theta rhythms exist in normal infants and children as well as during drowsiness and sleep in adults. Only a small amount of theta rhythms appears in the normal waking adult. Presence of high theta activity in awake adults suggests abnormal and pathological conditions.

- *Alpha* (8-12 Hz): Alpha rhythms exist in normal adults during relaxed and mentally inactive awakeness. The amplitude is mostly less than 50μV and appears most prominently in the occipital region. Alpha

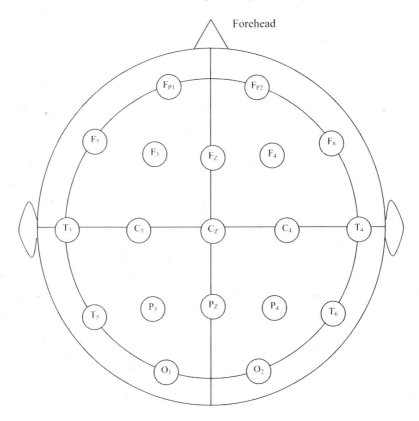

FIGURE 5.2
Standard international 10-20 configuration [adapted from American Clinical
Neurophysiology Society (2006)]

rhythms are blocked by opening of the eyes (visual attention) and other
mental efforts such as thinking.

- *Beta* (13-30 Hz): Beta rhythms are primarily found in the frontocentral
  regions with lower amplitude than alpha rhythms. They are enhanced
  by expectancy states and tension.

- *Gamma* (>30 Hz): Gamma rhythms, the high frequency band, are usually not of much clinical and physiological interest and therefore often
  filtered out in EEG recordings.

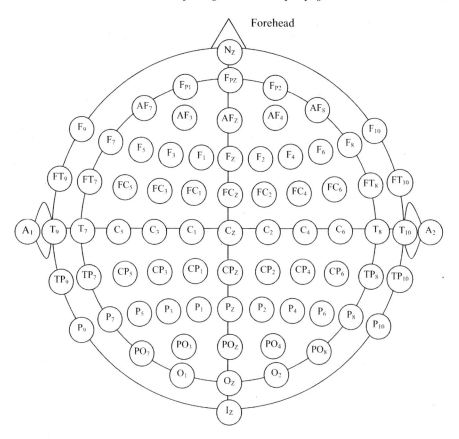

FIGURE 5.3
Modified and expanded 10-10 configuration [adapted from American Clinical
Neurophysiology Society (2006)]

EEGs contain a wealth of information which can be mined to yield tremen-
dous insight into the dynamics of the human brain. Conventional visual inspection of EEGs by trained neurologists includes the examination of the following
features: frequency or wavelength, voltage or amplitude, waveform regularity,
and reactivity to eye opening, hyperventilation, and photic stimulation. It also
includes the identification of the spatial range (local or generalized, unilateral
or bilateral) and temporal persistence (sporadic and brief or prolonged and
persistent) of the abnormalities. Although such features may be detected from

a visual inspection by trained neurologists, automated algorithms in a clinical setting do not fare well.

Moreover, other information such as complicated hidden patterns underlying the EEG waveforms is *invisible* and has to be extracted using advanced mathematical and analytical tools. This is mostly uncharted territory and many such invisible sources of information remain undiscovered. In recent years, various paradigms have been applied to overcome these problems, identify new markers of abnormality, and automate the identification of markers in EEG analysis (Iasemidis and Sackellares, 1991; Elger and Lehnertz, 1994, 1998; Iasemidis et al., 1994, 2003; Adeli et al., 2003).

5.3   Data Mining Techniques

Although EEGs contain a vast amount of information, not all EEG components are useful. Very high frequency fluctuations can be attributed to electromagnetic interference. Artifacts in various frequency ranges may be generated due to electrical events such as muscle movements, eye blinks, and heart beats, to name a few. These components of the EEG can be characterized as noise and need to be discarded. Signal analysis and processing techniques such as time-frequency analysis and wavelet transforms are used to extract relevant information.

Denoising the EEG using time-frequency and wavelet analysis attempts to yield a *clean* signal which needs to be mathematically analyzed in order to obtain features or markers of abnormality that can distinguish between normal and abnormal states. Recent research has demonstrated that a spatio-temporal investigation of the underlying non-linear chaotic dynamics of EEGs can yield such markers. To discover the chaotic dynamics underlying the EEG, studies have been performed, with some success, on EEGs obtained from both (a) normal states of the brain such as sleep (Molnar and Skinner, 1991; Roschke and Aldenhoff, 1991; Niestroj et al., 1995; Zhang et al., 2001; Kobayashi et al., 2001, 2002; Ferri et al., 1998, 2002, 2003; Shen et al., 2003) and meditation (Aftanas and Golocheikine, 2002; Efremova and Kulikov, 2002), as well as (b) pathological states such as schizophrenia (Roschke and Aldenhoff, 1993; Paulus et al., 1996; Huber et al., 1999, 2000; Paulus and Braff, 2003) and epilepsy (Iasemidis and Sackellares, 1991; Bullmore et al., 1992; Iasemidis et al., 1994; Lopes da Silva et al., 1994; Elger and Lehnertz, 1994, 1998; Hively et al., 1999; Andrzejak et al., 2001; Litt and Echauz, 2002; Adeli et al., 2003, 2007; Ghosh-Dastidar et al., 2007).

Although the wavelet transform can be used for denoising, in this book it also forms the basis for a novel integrated wavelet-chaos methodology. This methodology challenges the assumption that the EEG represents the dynamics of the entire brain as a unified system and needs to be treated as a whole for investigation of the chaotic dynamics. Chaos and wavelet analyses are adroitly integrated to identify features that best characterize the state of the brain. These features can be used to study differences in various brain states.

Due to the wide variety of features, the problem of EEG-based detection and diagnosis of brain states and disorders may be approached as pattern recognition problems where the pattern to be recognized or classified may be spatial, temporal, or spatio-temporal sequences in these feature spaces. Various methods such as cluster analysis, discriminant analysis, and even artificial neural networks (ANNs) can be employed for this purpose. Besthorn et al. (1997) report that linear discriminant analysis of EEGs yields more accurate classification compared with k-means cluster analysis.

ANNs have evolved as a powerful tool for pattern recognition, classification, prediction, and pattern completion. These networks are trained using real-world data and can provide approximate solutions to problems that are not easily solved by conventional mathematical approaches. Studies have reported that ANNs yield greater classification accuracy than traditional methods such as statistical, clustering, and discriminant analysis (Anderer et al., 1994; Besthorn et al., 1997). ANN classification accuracy is further increased if the EEGs are preprocessed with chaos analysis (Pritchard et al., 1994) or wavelet analysis (Polikar et al., 1997; Petrosian et al., 2000a, 2001).

Chaos theory, time-frequency and wavelet analysis, and artificial neural networks are computational tools that can be applied in complementary roles toward the common goal of analysis of EEG waveforms. The application depends on how the solution to the problem is conceptualized. For instance, in the case of detecting EEG abnormalities, the problem can be approached as (1) searching for an abnormal waveform using statistical similarities with known waveforms such as a wavelet, (2) recognizing abnormal changes in the EEG patterns using an artificial neural network, (3) quantifying changes in underlying non-linear chaotic EEG dynamics and using the quantifier as a marker of abnormality, or in the case of a particularly difficult problem, (4) a judicious combination of the above. Another instance of using the multi-paradigm approach is when one tool is designed so as to enhance the performance of another. For example, the time-frequency or wavelet analysis is employed for the extraction of a feature that may be classified more accurately by a classification algorithm. In this book, the complementary roles of these methods are investigated toward the long term goal of increasing the accuracy and performance of automated detection and diagnosis algorithms.

5.4   Multi-Paradigm Data Mining Strategy for EEGs

A novel multi-paradigm computational model for data mining and pattern recognition of EEGs requires an adroit integration of chaos analysis, wavelets, and artificial neural networks. In this book, various methodologies are explored and applied to EEGs from subjects with two types of seizures: absence seizures and seizures resulting from temporal lobe epilepsy (TLE). These two types of seizures are fundamentally different from each other. The former is associated with visually recognizable epileptic waveforms such as the *3-Hz spike and slow wave complex*. The latter is also associated with abnormal waveforms but these waveforms cannot be characterized easily.

One goal is to assess the applicability of wavelet transforms for characterization of epileptic waveforms such as the 3-Hz spike and slow wave complex observed in absence seizures as presented in Chapter 6. Another goal is to discover EEG markers of abnormality in TLE that cannot ordinarily be obtained from a visual EEG inspection as presented in Chapters 7 to 9. Toward the second goal, the temporal evolution of the brain state (as represented by the EEG) in these disorders is investigated. Subsequently, a judicious combination of various invisible as well as visible EEG markers is presented as a basis for development of new in vivo automated detection and diagnosis tools.

Epilepsy diagnosis and seizure detection are modeled as a clinically significant EEG classification problem. An automated computer model that can accurately differentiate between normal EEGs from interictal EEGs can be used to diagnose epilepsy in a clinical setting. A model that can accurately differentiate between interictal and ictal EEGs can be used to detect seizures in the environment of epilepsy monitoring units. In a clinical setting, the distinctions between these different EEG groups are often not very well defined.

Therefore, for real-time applicability, it is imperative that the model be able to identify EEGs obtained under the above mentioned conditions accurately and consistently.

The goal of consistent and accurate classification is approached from two different angles: 1) designing an appropriate feature space by identifying combinations of parameters that increase the inter-class separation and 2) designing a classifier that can accurately model the classification problem based on the selected feature space. Therefore, corresponding to the above mentioned two angles, the research presented in this book is organized into the following two phases:

1. Feature space identification and feature enhancement using wavelet-chaos methodology

2. Development of accurate and robust classifiers

5.4.1   Feature Space Identification and Feature Enhancement Using Wavelet-Chaos Methodology

The first phase, as presented in Chapters 7 and 8, is focused on identifying a feature space that can maximize the separation between normal and abnormal EEGs. The feature space consists of various features, defined as parameters or waveforms in EEGs that may be used as markers of abnormality for diagnosis of TLE. Visible features such as focal spikes can be characterized by specific wavelets or other waveforms. Alternatively, certain features may be characteristics of the underlying chaotic dynamics and therefore not detectable from a visual inspection of the EEG.

The key to maximizing the usefulness of any feature is to maximize its *detectability* (consistency and accuracy of detection). This is achieved in two ways. One, computational methods are developed to amplify the features to be detected. Two, new features/markers are discovered that are inherently *better* in terms of detectability. The objective is to develop new features by means of quantification of differences in normal and abnormal chaotic dynamics underlying the EEG in TLE. This phase is organized into the following tasks:

1. Denoising and artifact removal: EEG signals often contain various types of artifacts (especially myogenic artifacts) and electromagnetic interference which may lead to inaccurate analysis by masking transients that may be predictive of the disorder under study. Filters based on time-frequency and wavelet analyses are used to remove or minimize the presence of these artifacts to allow for more effective feature identification.

2. Sub-band analysis: Changes in EEG dynamics may exist in certain frequency ranges but not others. Moreover, these changes may not show up in the complete EEG. In this analysis methodology, the EEG is decomposed into its physiological alpha, beta, gamma, theta, and delta sub-bands using a wavelet-based approach. The methodology is investigated with strategically selected frequency bands to isolate deterministic chaos and other changes in each frequency band.

3. Chaos quantification: Parameters quantifying chaotic dynamics in EEG time series include the correlation dimension (CD) and largest Lyapunov exponent (LLE). Various combinations of these parameters obtained from the entire EEG and EEG sub-bands are investigated as possible feature spaces for accurate EEG representation.

4. Statistical analysis and the mixed-band feature space identification: Parametric and non-parametric analysis of variance (ANOVA) tests (as determined by the data distribution) are performed in order to assess the statistical significance of the differences in the identified features. On the basis of the statistical analysis, mixed-band feature spaces based

on various EEG sub-bands are identified with the goal of increasing the separation between normal and abnormal EEGs.

The wavelet-chaos methodology performs two primary functions:

1. Increasing the classification accuracy: The parameters obtained from the wavelet-chaos analysis are based on chaotic nonlinear dynamics of the brain. An extensive mixed-band analysis involving various combinations of features from various sub-bands is performed in order to identify features that are significantly different across the EEG groups.

2. Data reduction: The methodology maps each EEG dataset to a point on the P-dimensional feature space with each selected parameter as a coordinate axis, where P is the number of selected features.

5.4.2   Development of Accurate and Robust Classifiers

In the second phase, described in Chapters 8 and 9, new classification models that maximize the classification accuracy and robustness are presented. The classification strategies are presented as follows:

1. Evaluation of the mixed-band feature space: Combinations of various wavelet-chaos features and classifiers such as k-means clustering, linear and quadratic discriminant analyses, and backpropagation and radial basis function neural networks are investigated to determine the combination that consistently yields the highest classification accuracy. This provides a benchmark for evaluating the performance of new improved neural network classifiers.

2. Development of improved radial basis function neural network classifier: A new radial basis function neural network (RBFNN) classifier is presented to increase the robustness of classification.

3. Development of spiking neural network: Supervised learning using a spiking neural network (SNN) is a new development in the field of artificial intelligence and is in its infancy. Only simplistic models have been developed that have been used to solve very simple classification problems such as the XOR problem. The SNN, however, has great potential because it mimics the dynamics of the biological neuron in greater detail. The goal is to develop more powerful models of learning which could lead to more accurate and robust classifiers. In this book, a new SNN model and supervised learning algorithm are presented and applied to the complicated EEG classification problem. Spiking neural networks are, in themselves, a separate research topic and are therefore presented separately in Part IV.

5.5 Epilepsy and Epileptic Seizures

Epilepsy is a common brain disorder that affects about 1% of the population in the United States and is characterized by intermittent abnormal firing of neurons in the brain which may lead to recurrent and spontaneous seizures (with no apparent external cause or trigger). Approximately 30% of the epileptic population is not helped by medications (Porter, 1993; National Institute of Neurological Disorders and Stroke, 2004). Seizures can be categorized as generalized or partial depending on established conventions as explained below. Generalized seizures occur due to simultaneous abnormal activity in multiple parts of both brain hemispheres from the beginning leading to tonic-clonic activity and loss of consciousness. Partial (or focal) seizures are more common and initiate in one part of the brain, often leading to strange sensations, motor behavior, and even loss of memory. These seizures are further subdivided based on the part of the brain that contains the epileptogenic focus which determines the exact symptoms. Partial seizures can sometimes spread from the focus to other parts of the brain, leading to secondary generalized seizures.

Different parts of the brain are implicated in the generation of different types of seizures associated with various types of epilepsy. For instance, primary generalized seizures such as absence seizures are attributed to generally increased cortical excitability and disturbances in the thalamocortical pathways (Westbrook, 2000). On the other hand, partial seizures are attributed to localized disturbances in various areas of the brain. Due to this reason, there is no one area of the brain that can be implicated in the generation of all epilepsies. However, in almost 33% of all patients with partial seizures, the epileptogenic focus (or at least one of the foci) is located in the temporal lobe (Devinsky, 2004). This condition is termed temporal lobe epilepsy. TLE seizures are of primary clinical importance due to the frequency of occurrence and difficulty of diagnosis and treatment.

TLE can be further categorized as mesial or neocortical based on the location of the focus inside the temporal lobe. In mesial TLE, which is more frequent, the epileptogenic focus is located in the mesial temporal lobe which consists of the amygdala, hippocampal formation (HF), and the parahippocampal gyrus. Out of these, the most common location of the focus is the HF and, in some cases, the amygdala (Najm et al., 2001). In neocortical TLE, the focus exists in the neocortical (lateral) temporal lobe. Due to the close proximity of the two areas and extensive interconnectivity, abnormal neuronal discharges spread easily from one area to the other, making it very difficult to distinguish between the two types and their symptoms (Kotagal, 2001). The HF is implicated as the epileptogenic focus more frequently and therefore appears to be more important than other areas in epilepsy.

Epilepsy is often described as a group of disorders with many types, sub-types, and cross-classifications. The two types presented as example applications in this book have been selected because of their clinical relevance. Absence (*petit mal*) seizure is one of the main types of generalized seizures and the underlying pathophysiology is not completely understood. The diagnosis is made by neurologists primarily based on a visual identification of the 3-Hz spike and slow wave complex. The significance of TLE is even greater due to its predominance, difficulty of diagnosis, and significant personal and societal impact. As a result, most of this part of the book is devoted to application of the wavelet-chaos-neural network methodologies to TLE diagnosis and seizure detection. The presented methodologies, however, are certainly not limited to these two types of epilepsy. Rather, they are applicable to neurological disorders in general. To illustrate this point, we describe a possible application of the wavelet-chaos methodology to Alzheimer's disease in Part III.

6

Analysis of EEGs in an Epileptic Patient Using Wavelet Transform

6.1 Introduction

Epileptic seizures manifest themselves as abnormalities in electroencephalograms (EEGs) and are characterized by brief and episodic neuronal synchronous discharges with dramatically increased amplitude. In partial seizures, this anomalous synchrony occurs in the brain locally and is observed only in a few channels of the EEG. In generalized seizures the entire brain is involved and the discharges can be observed in every channel of the EEG. Wavelet transforms are an effective time-frequency analysis tool for analyzing such transient non-stationary signals. In this chapter, the applicability of wavelet transforms is demonstrated for the analysis of epileptiform discharges in EEGs of subjects with absence seizure. The excellent feature extraction and representation properties of wavelet transforms are used to analyze individual transient events such as the 3-Hz spike and slow wave complex.

Four channels of an EEG, F7-C3, F8-C4, T5-O1, and T6-O2 (Fig. 5.2), recorded from a patient with two absence seizure epileptic discharges are

This chapter is based on the article: Adeli, H., Zhou, Z., and Dadmehr, N. (2003), "Analysis of EEG Records in an Epileptic Patient Using Wavelet Transform", Journal of Neuroscience Methods, 123(1), pp. 69-87, and is reproduced by the permission of the publisher, Elsevier.

shown in Figs. 6.1 and 6.2 (data available online at: `ftp://sigftp.cs.tut.fi/pub/eeg-data/tv0003b0.rec` and `tv0004b0.rec`). It should be noted that the four channels shown are actually the mathematical difference in the signals between pairs of electrodes. For instance, the F7-C3 channel represents the C3 signal subtracted from the F7 signal. These modified channel configurations are called *montages* and are often used to highlight location-specific changes in the EEG and minimize noise. The signals show the 3-Hz spike and slow wave complex during the time period 3.5-7 seconds. Although these two seizures may look similar in terms of displaying the raw EEG data, they may have different underlying time-frequency structures not readily discernable from the raw data by visual inspection. As such, both are analyzed in this chapter.

Currently, the analysis of the EEG data is performed primarily by neurologists through visual inspection. Fourier analysis has also been used to analyze EEGs (Gotman, 1990). Most studies on the characteristics of the 3-Hz spike and slow wave complex have been based on simple visual inspection of data recorded for different channels.

After visual inspection of the EEGs with absence seizure discharges, Weir (1965) pointed out that a surface negative spike followed by a surface negative slow wave oversimplifies the 3-Hz spike and slow wave complexes. He observed four components including an initial surface positive transient wave followed by a low voltage (25-50 μV) surface negative spike and a *classical* surface negative spike lasting 4-60 ms. The first brief spike appears in the centrotemporal regions and the second in frontal areas. The fourth and final surface negative wave is in conjunction with the declining initial positive transient wave. In other words, there are two spikes between the first surface positive transient wave and the last surface negative wave. Rodin and Ancheta (1987) analyzed the 3-Hz spike and slow wave complex for five patients using com-

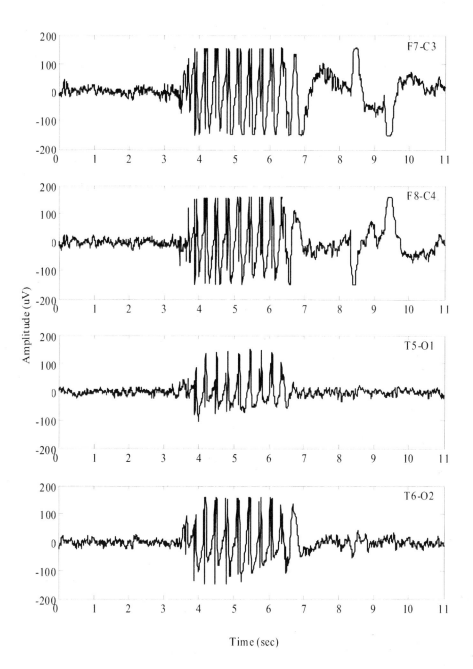

FIGURE 6.1
EEG segment showing epileptic discharges during an absence seizure

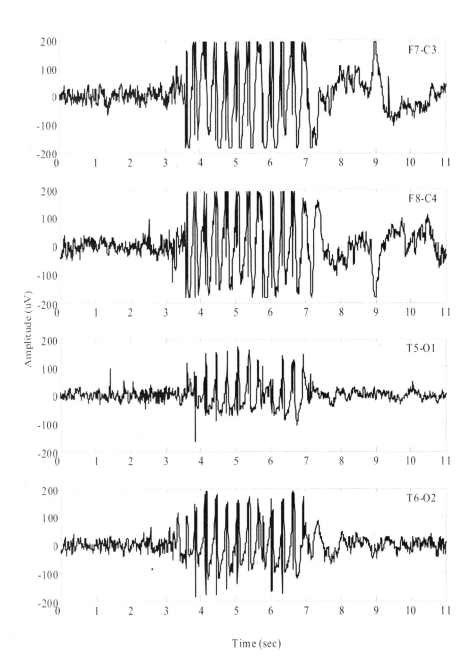

FIGURE 6.2
EEG segment showing epileptic discharges during a second absence seizure

puterized topographic mapping and concluded that the maximal amplitudes occur in frontal regions. They also observed that the occurrence of spikes in one hemisphere randomly precedes the other by several milliseconds.

6.2   Wavelet Analysis of a Normal EEG

In this section, DWT is applied to part of a normal EEG and the results are interpreted. The EEG was obtained from an online database (`http://kdd.`
`ics.uci.edu/databases/eeg/eeg.html`). It was recorded from 64 electrodes placed on the scalp of a normal adult (representing 32 EEG channels, since each channel represents the potential difference between pairs of electrodes) who was exposed to one visual stimulus (pictures of objects). The EEG signal was sampled at 256 Hz (sampling interval of $1/256 = 0.0039$ seconds or 3.9 milliseconds). A one-second EEG for one of the channels is shown at the top of the left column in Fig. 6.3 (identified by the letter **s**).

Third order Daubechies wavelet transform was applied to the EEG signal. The results are shown in Fig. 6.3 with five different levels of approximation (identified by **a1** to **a5** and displayed in the left column) and details (identified by **d1** to **d5** and displayed in the right-hand column). These approximation and detail components of the EEG are reconstructed from the wavelet coefficients. Approximation **a4** is obtained by superimposing details **d5** on approximation **a5**. Approximation **a3** is obtained by superimposing details **d4** on approximation **a4**, and so on. Finally, the original signal **s** is obtained by superimposing details **d1** on approximation **a1**. The utility of the wavelet transform as a mathematical microscope is clear. For example, the approximations at levels 3 and 4 show good overall trend information for the signal, which is primarily due to an eye blink artifact, with the higher frequency in-

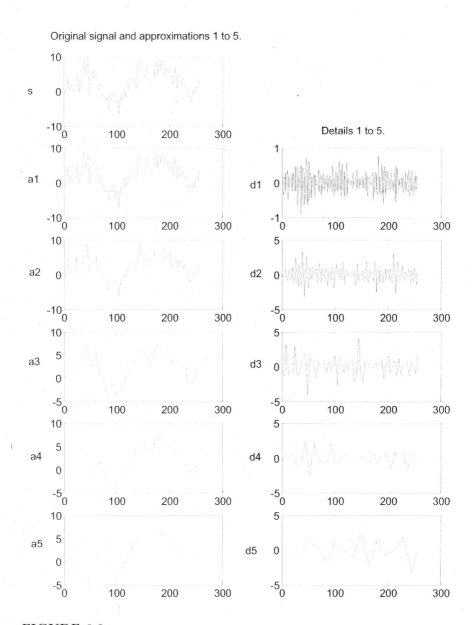

FIGURE 6.3
Third order Daubechies wavelet transform of a one-second EEG segment recorded from a normal adult

formation retained in the details. Additional details are presented in the lower scale levels 2 and 1.

Table 6.1 presents frequencies and periods for various levels of decomposition for the third order Daubechies wavelet with a sampling frequency of 256 Hz. It can be seen from Table 6.1 that the components from level 5 decomposition are approximately within the *theta* range (4-7 Hz), those from level 4 decomposition are within the *alpha* range (8-12 Hz), and those from level 3 decomposition are within the *beta* range (13-30 Hz). Lower level decompositions corresponding to higher frequencies have negligible magnitudes in a normal EEG. The corresponding frequencies and periods for various levels of decomposition for the third order Daubechies wavelet with a sampling frequency of 200 Hz are summarized in Table 6.2.

TABLE 6.1
Frequencies and periods for various decomposition levels for the third order Daubechies wavelet (sampling frequency of 256 Hz)

Level of decomposition (i)	0	1	2	3	4	5	6
Scale (2^i)	1	2	4	8	16	32	64
Frequency (Hz)	204.8	102.4	51.2	25.6	12.8	6.4	3.2
Period (sec)	0.0049	0.098	0.0195	0.0391	0.0781	0.1563	0.3125

TABLE 6.2
Frequencies and periods for various decomposition levels for the third order Daubechies wavelet (sampling frequency of 200 Hz)

Level of decomposition (i)	0	1	2	3	4	5	6
Scale (2^i)	1	2	4	8	16	32	64
Frequency (Hz)	142.86	71.429	35.715	17.857	8.929	4.464	2.232
Period (sec)	0.007	0.014	0.028	0.056	0.112	0.224	0.448

6.3   Characterization of the 3-Hz Spike and Slow Wave Complex in Absence Seizures Using Wavelet Transforms

In contrast to Fourier analysis, where only sinusoid functions are employed as a basis for analysis, there exist families of functions that serve as wavelet basis functions. Identifying and selecting the most appropriate wavelet basis function for a given signal analysis problem should be properly investigated. Issues to be considered in the selection process include interpretation of the transformed data, desired level of resolution, accuracy, and computational efficiency. In this chapter, Daubechies and harmonic wavelets are investigated for the analysis of epileptic EEGs.

6.3.1   Daubechies Wavelets

Daubechies wavelets (Daubechies, 1988) of different orders (2, 3, 4, 5, and 6) were investigated for the analysis of epileptic EEGs. This family of wavelets is known for its orthogonality property and efficient filter implementation. The fourth order Daubechies wavelet was found to be the most appropriate for analysis of epileptic EEG data. The lower order wavelets of the family were found to be too coarse to represent EEG spikes properly. The higher order ones have more oscillations and cannot accurately characterize the spiky form of the absence seizure epileptic EEG investigated in this chapter.

Figures 6.4 to 6.11 show the single channel EEGs containing absence seizure epileptic discharges shown in Figs. 6.1 and 6.2 and their fourth order Daubechies wavelet approximations and details. In these figures, it should be noted that for the sake of displaying the details the vertical scales are different for different levels of decomposition. Six different levels of approximations (**a1** to **a6**) and details (**d1** to **d6**) are presented in the left and right columns, respectively. The EEG is sampled at 200 Hz (200 data points per second) to yield 2048 data points. In Figs. 6.4 to 6.7, the seizure starts at approximately the 750th point and ends at the 1300th point. In Figs. 6.8 to 6.11, the seizure starts at approximately the 700th point and ends at approximately the 1400th point. The original signal **s** is the sum of the first level of approximation **a1** and the first level of detail **d1**; the first level approximation **a1** is the sum of the second level approximation **a2** and the second level detail **d2**; and so on. Therefore, high frequency components are stripped from the signal approximation layer by layer. By inspecting the details at each level, the time-frequency feature of the original signal is obtained. A close inspection of Figs. 6.4 to 6.11 indicates that the spike and wave trains manifested in the EEGs are captured accurately in the transformed detail signals **d5** and **d3**. The former represents the epileptic slow wave with high amplitude and the latter represents spikes effectively.

6.3.2 Harmonic Wavelet

Harmonic wavelet functions are suitable for analysis and characterization of the EEGs because they can be used to precisely locate the frequency bands of interest. The same EEG signals presented in Figs. 6.1 and 6.2 are analyzed using the discrete harmonic wavelet transform. The resulting wavelet coefficient moduli or signal energies for six different resolution levels ($j = 1$ to 6) are plotted in Figs. 6.12 to 6.19. In these figures also the vertical scales are different for different levels of decomposition. As can be seen in these figures and Table 6.3, the frequency range halves with each increment in the decomposition level. In the absence seizure spike and wave complexes, most of the energy of the spike is concentrated in level $j = 3$ with a frequency range of 12.5-25 Hz indicating the width of the spike in the frequency domain, while

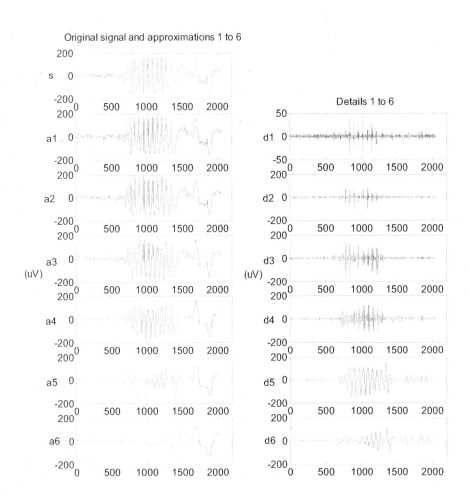

FIGURE 6.4
Fourth order Daubechies wavelet transform of the F7-C3 EEG channel for the
absence seizure 1 shown in Fig. 6.1

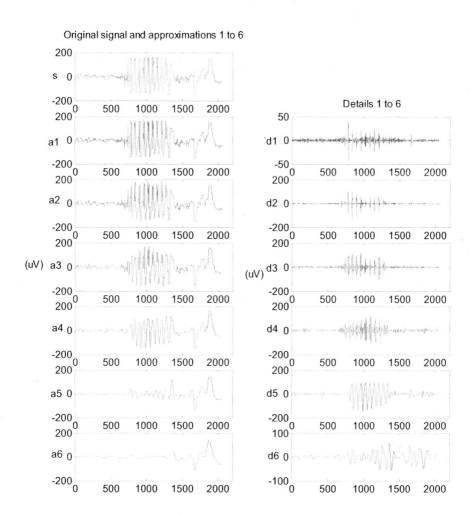

FIGURE 6.5
Fourth order Daubechies wavelet transform of the F8-C4 EEG channel for the absence seizure 1 shown in Fig. 6.1

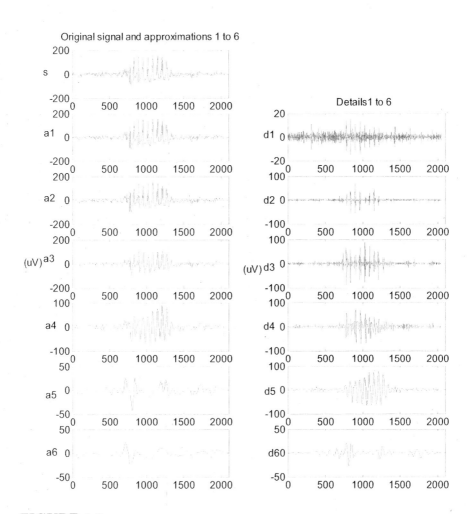

FIGURE 6.6
Fourth order Daubechies wavelet transform of the T5-O1 EEG channel for
the absence seizure 1 shown in Fig. 6.1

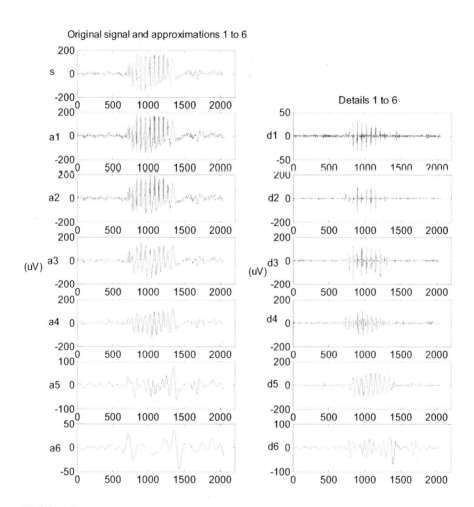

FIGURE 6.7
Fourth order Daubechies wavelet transform of the T6-O2 EEG channel for
the absence seizure 1 shown in Fig. 6.1

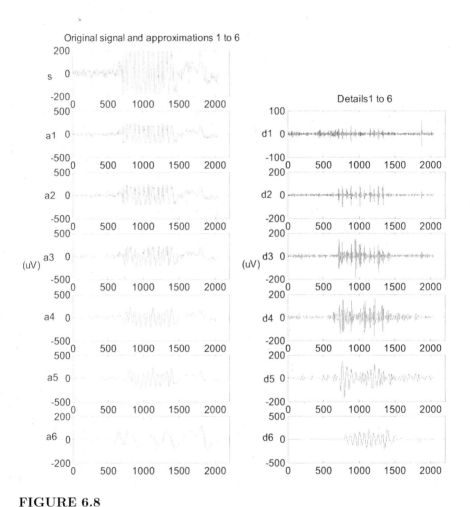

FIGURE 6.8

Fourth order Daubechies wavelet transform of the F7-C3 EEG channel for the absence seizure 2 shown in Fig. 6.2

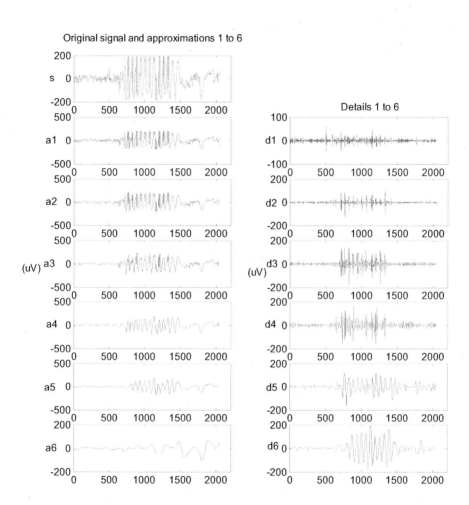

FIGURE 6.9
Fourth order Daubechies wavelet transform of the F8-C4 EEG channel for the absence seizure 2 shown in Fig. 6.2

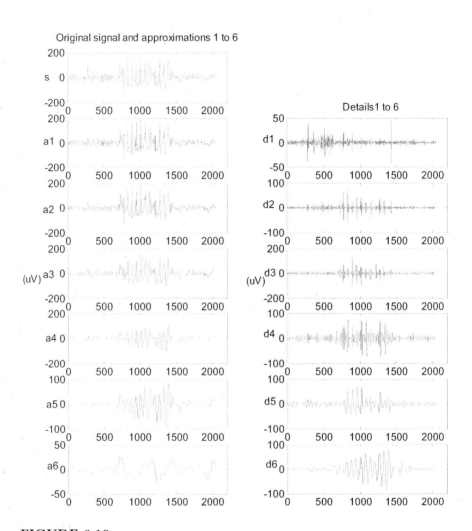

FIGURE 6.10
Fourth order Daubechies wavelet transform of the T5-O1 EEG channel for the absence seizure 2 shown in Fig. 6.2

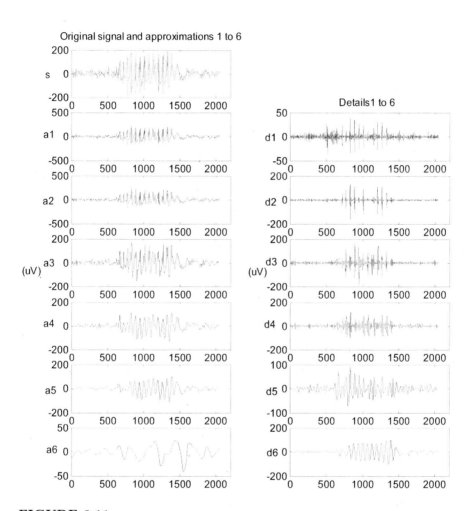

FIGURE 6.11

Fourth order Daubechies wavelet transform of the T6-O2 EEG channel for the absence seizure 2 shown in Fig. 6.2

most of the slow wave energy is concentrated in level $j = 5$ with a frequency range of 3.125-6.25 Hz indicating the frequency band of the low wave in the frequency domain.

TABLE 6.3
Frequency ranges for various decomposition levels for harmonic wavelet (sampling frequency of 200 Hz)

Level of decomposition (i)	1	2	3	4	5	6
Scale (2^i)	2	4	8	16	32	64
Frequency (Hz)	50-100	25-50	12.5-25	6.25-12.5	3.125-6.25	1.5625-3.125

6.3.3   Characterization

Based on the examination of the results for both fourth order Daubechies and harmonic wavelet decompositions of EEGs from a subject with absence seizure, the following observations are made:

- Both high and low frequency components have greater amplitudes in the frontal region (Figs. 6.4, 6.5, 6.8, and 6.9) than those for the occipital region (Figs. 6.6, 6.7, 6.10, and 6.11).

- In frontal regions:

  - High frequency oscillations appear in the early stage of the epileptic discharge with an amplitude of about 1/4 of the 3-Hz spike and slow wave complex, as demonstrated in the level 1 detail signals **d1** in Figs. 6.4 and 6.5.

  - Low frequency waves occur in later stages of the same spike and wave complex as can be seen in the level 6 detail signal **d6** in Figs. 6.4 and 6.5.

  - The same observations are made in Figs. 6.12, 6.13, 6.16, and 6.17

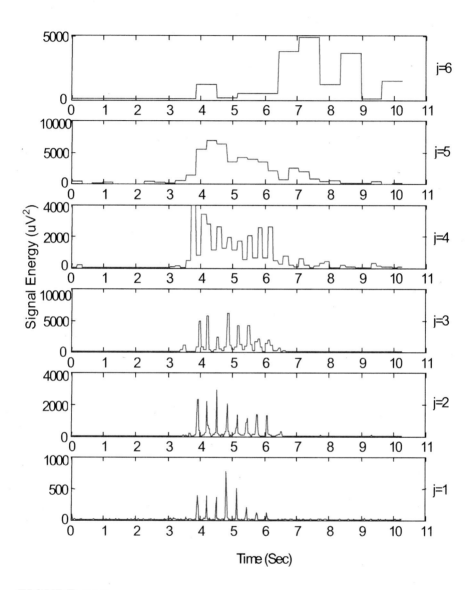

FIGURE 6.12
Harmonic wavelet transform of the F7-C3 EEG channel for the absence seizure 1 shown in Fig. 6.1

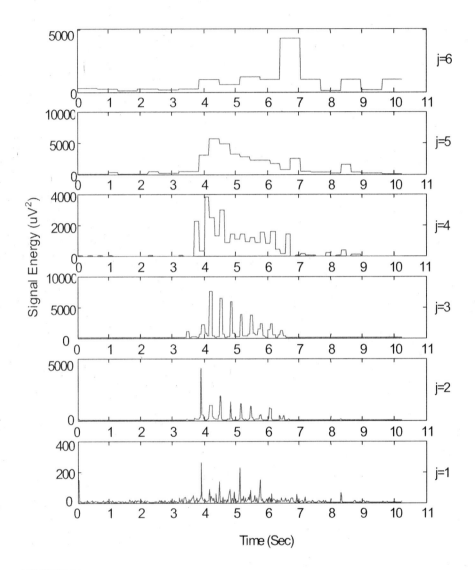

FIGURE 6.13

Harmonic wavelet transform of the F8-C4 EEG channel for the absence seizure 1 shown in Fig. 6.1

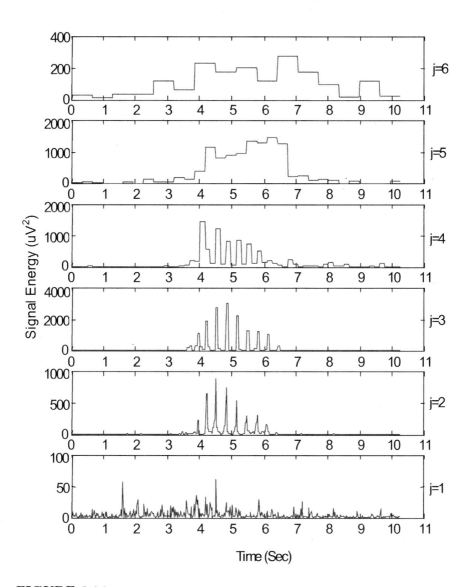

FIGURE 6.14
Harmonic wavelet transform of the T5-O1 EEG channel for the absence seizure 1 shown in Fig. 6.1

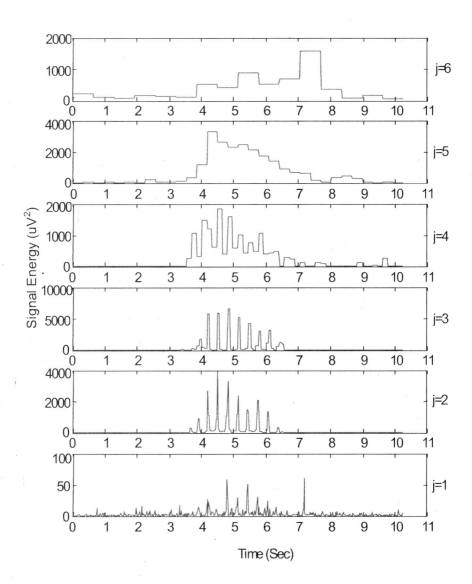

FIGURE 6.15
Harmonic wavelet transform of the T6-O2 EEG channel for the absence seizure
1 shown in Fig. 6.1

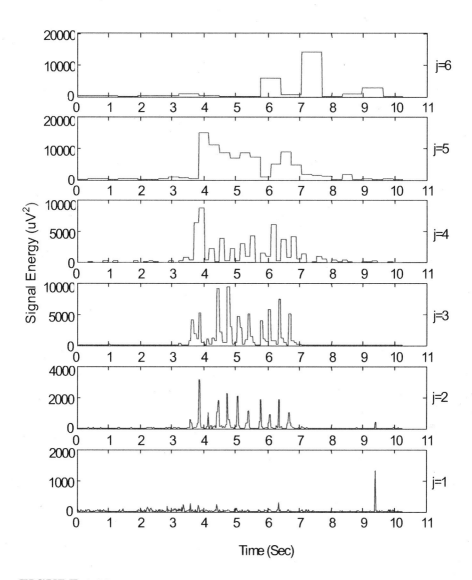

FIGURE 6.16
Harmonic wavelet transform of the F7-C3 EEG channel for the absence seizure 2 shown in Fig. 6.2

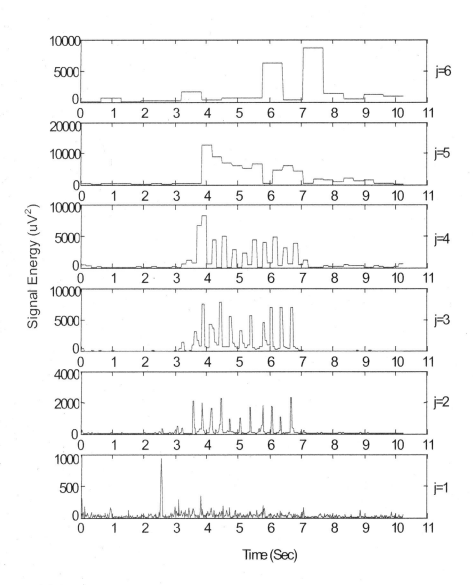

FIGURE 6.17
Harmonic wavelet transform of the F8-C4 EEG channel for the absence seizure
2 shown in Fig. 6.2

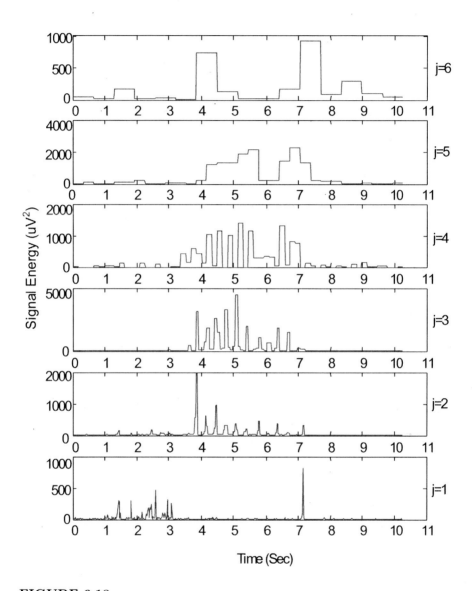

FIGURE 6.18

Harmonic wavelet transform of the T5-O1 EEG channel for the absence seizure 2 shown in Fig. 6.2'

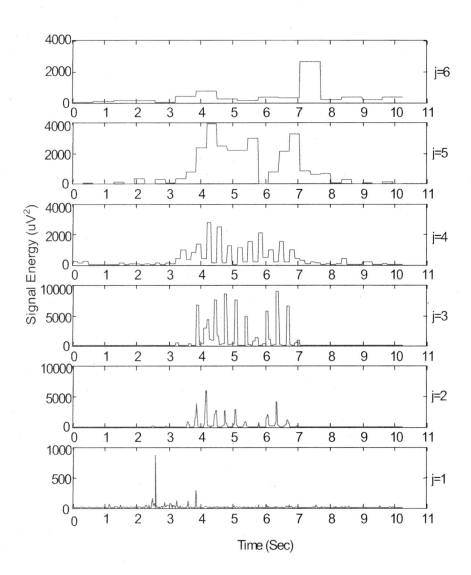

FIGURE 6.19

Harmonic wavelet transform of the T6-O2 EEG channel for the absence seizure
2 shown in Fig. 6.2

by comparing the resolution level $j = 6$ signal (lower frequency, in the delta range) with the level $j = 5$ signal (higher frequency, in the theta range).

- In occipital regions:

  - High amplitudes of low frequency transients are seen at both the beginning and end of the first seizure (Fig. 6.6 **d6**, Fig. 6.7 **a6**). In the second seizure, however, high amplitudes of low frequency transients are seen at the beginning, in the middle, and at the end of the seizure (Figs. 6.10, 6.11 **a6**) suggesting that the second seizure is a summation of two successive shorter-duration seizures.

  - Based on the fourth order Daubechies wavelet transform, high frequency components are more intense in the right hemisphere (Figs. 6.7 **d1** and 6.11 **d1**) than in the left one (Figs. 6.6 **d2** and 6.10 **d2**). The same observation is made using harmonic wavelets at decomposition level $j = 3$ (frequency range of 12.5 to 25 Hz) when Figs. 6.14 and 6.18 are compared with Figs. 6.15 and 6.19, respectively.

- The frequency components of the signal in the ultra high frequency (50-100 Hz) range (level $j = 1$ in Figs. 6.12 to 6.19) are low in amplitude and can be ignored in clinical analysis. In fact, this frequency range is usually filtered out by a high-pass filter when the signal is originally sampled. The appearance of some signal in this frequency range may be attributed to noise during recording.

- The two different spikes observed by Weir (1965), denoted as *classical* and *secondary*, are demonstrated separately in harmonic wavelets. The classical spike is observed in level $j = 3$; the secondary spike with a smaller amplitude appears earlier in time and is observed in level $j = 2$ (Figs. 6.12 to 6.19).

- An overall inspection of harmonic wavelets in Figs. 6.12 to 6.19 shows
  that the shape and trend at higher levels of signal energy distribution
  (levels 4, 5 and 6, with lower frequency content) over time precisely
  echo the shape and trend in lower levels of signal energy distribution
  (levels 2 and 3, with higher frequency content) suggesting a synchronous
  firing of neurons during the seizure at different scales, but first at a high
  frequency range followed by a lower frequency range.

6.4   Concluding Remarks

Through wavelet decomposition of the 3-Hz spike and slow wave epileptic discharges in EEGs, transient features are accurately captured and localized in
both time and frequency context. The capability of this mathematical microscope to analyze different scales of neural rhythms is shown to be a powerful
tool for investigating small-scale oscillations of the brain signals. However,
to utilize this mathematical microscope effectively, the best suitable wavelet
basis function has to be identified for the particular application. Fourth order Daubechies and harmonic wavelets are experimentally found to be very
appropriate for analysis of the 3-Hz spike and slow wave complex in EEGs.
Wavelet analyses of EEGs obtained from a population of epileptic subjects can
potentially provide deeper insight into the physiological processes underlying
brain dynamics at seizure onset.

In the next few chapters, computational models will be presented for automated detection of epileptic discharges in other types of seizures that may,
in the future, be used to predict the onset of seizure. Accurate detection
of various types of seizure is a complicated problem requiring analysis of a
large set of EEGs. The final objective is to create a system for continuous

EEG-based monitoring of hospitalized patients and an automated detection or early-warning system for epileptic episodes. Such a system can also be used for automated administration of medication through implantable drug delivery systems.

7

Wavelet-Chaos Methodology for Analysis of EEGs and EEG Sub-Bands

---

7.1 Introduction

Temporal lobe epilepsy (TLE) is characterized by intermittent abnormal neuronal firing in the brain which can lead to seizures. Ictal brain activity (during a seizure) differs significantly from the activity in the normal state with respect to both frequency as well as pattern of neuronal firing. In contrast to normal brain activity in which neurons fire about 80 times per second, neurons may fire as fast as 500 times per second during a seizure. Further, the spatio-temporal pattern of neuronal firing gradually evolves from a normal state, first to a preictal (interictal) state and then to an ictal state (Iasemidis et al., 1994; Lopes da Silva et al., 1994).

Normally, neurons in different parts of the brain fire independently of each other. In contrast, in the interictal state, neurons start firing in multiple parts of the brain in synchronization with the epileptogenic focus which leads to the ictal state. Despite these differences, detection of seizures can be challenging even from a visual inspection of the EEG by a trained neurologist for a variety of reasons such as excessive presence of myogenic artifacts, interference, and overlapping symptomatology of various mental states. Prediction of seizures is even more challenging because there is very little confirmed knowledge of the exact mechanism responsible for the seizure. Overcoming these obstacles by means of effective and accurate algorithms for automatic seizure detection and prediction can have a far reaching impact on diagnosis and treatment of epilepsy.

In recent years, a few attempts have been reported on seizure detection and prediction from EEG analysis using two different approaches: (a) examination of the waveforms in the preictal EEG to find events (markers) or changes in neuronal activity such as spikes (Gotman, 1999; Adeli et al., 2003; Durka, 2003) which may be precursors to seizures and (b) analysis of the nonlinear spatio-temporal evolution of the EEG signals to find a governing rule as the system moves from a seizure-free to seizure state (Iasemidis et al., 1994). Some work has also been reported using artificial neural networks (Adeli and Hung, 1995) for seizure prediction with wavelet preprocessing (Petrosian et al., 2000b). This chapter focuses on the second approach using a combination of chaos theory and wavelet analysis for non-linear dynamic analysis of EEGs. The multi-paradigm approach presented in this chapter is based on an integrated approach and simultaneous application of both chaos theory and wavelets for EEG analysis.

The approach presented in this book challenges the assumption that the EEG represents the dynamics of the entire brain as a unified system and needs to be treated as a whole. On the contrary, an EEG is a signal that represents the effect of the superimposition of diverse processes in the brain. Until recently, very little research has been done to separately study the effects of these individual processes. Each EEG is commonly decomposed into five EEG sub-bands: *delta* (0-4 Hz), *theta* (4-7 Hz), *alpha* (8-12 Hz), *beta* (13-30 Hz), and *gamma* (30-60 Hz). There is no good reason why the entire EEG should be more representative of brain dynamics than the individual frequency sub-bands. In fact, the sub-bands may yield more accurate information about constituent neuronal activities underlying the EEG and, consequently, certain changes in the EEGs that are not evident in the original full-spectrum EEG may be amplified when each sub-band is analyzed separately. This is a fundamental premise of our approach.

Iasemidis and Sackellares (1991) were among the first to study the nonlinear dynamics of EEG data in patients with temporal lobe epilepsy and in their subsequent studies concluded that the chaos in the brain was reduced in the preictal phase (Iasemidis et al., 2000b, 2003). Bullmore et al. (1992) performed a quantitative comparison of EEGs corresponding to normal and epileptic brain activity using fractal analysis. Similar findings confirming the reduction in complexity of neuronal firing during the preictal phase were also reported by Elger and Lehnertz (1994, 1998). Lopes da Silva et al. (1994) reported the synchronization of neuronal firing in different parts of the brain during a seizure. Hively et al. (1999) also proposed additional measures using chi-square statistics and phase space visitation frequency to quantify chaos in EEGs and to detect the transition from non-seizure to epileptic activity.

It has been reported that during normal brain activity, the pattern of neuronal firing represented by the EEG signal appears to be less organized and has greater complexity and chaoticity. However, prior to a seizure, the pattern of neuronal firing becomes more organized and is characterized by lower values of the largest Lyapunov exponent and correlation dimension of the chaotic attractor (Litt and Echauz, 2002; Iasemidis, 2003). There are conflicting reports on the values of the correlation dimension during a seizure. Hively et al. (1999) report values of correlation dimension greater than 6 during a seizure and between 1 and 2.6 otherwise, whereas Iasemidis et al. (2000b) report values in the range of 2 and 3 during a seizure.

A major shortcoming of existing seizure detection algorithms is their low accuracy resulting in high missed detection and false alarm rates. An inherent source of inaccuracy stems from the vast number of variables involved in the physiological system (Iasemidis, 2003). For instance, it cannot be decided if a false alarm is a true false alarm or just the precursor to a potential seizure that was prevented from occurring by changes that returned the neuronal pattern to the normal baseline. Another source of inaccuracy is that some of the parameters often used in the existing algorithms such as the correlation dimension and the largest Lyapunov exponent are obtained approximately from multidimensional experimental data.

A robust parameter capable of quantifying changes in the brain dynamics of epileptic patients, especially just before or during seizures, is key to accurate seizure detection and prediction. The robustness of this parameter is defined with respect to: 1) decreased sensitivity to physiological differences between individuals and inherent inaccuracies in EEGs such as noise and electrode artifacts, and 2) increased specificity to the particular disorder of interest, for example, temporal lobe epilepsy (TLE). Another impediment to accurate seizure detection is the lack of reliable standardized data. This reduces the statistical significance of the results since most of the EEG analysis reported in the literature is performed on a small number of data sets. In order to obtain a reliable estimate of the efficacy of the epilepsy detection parameters and algorithms, they should be tested on a relatively large number of data sets.

7.2   Wavelet-Chaos Analysis of EEG Signals

In this chapter, a wavelet-chaos methodology is presented for analysis of EEGs and EEG sub-bands for detection of seizure and epilepsy. It consists of three stages: a) wavelet analysis, b) preliminary chaos analysis, and c) final chaos analysis as outlined in Fig. 7.1. The methodology is applied to three different groups of EEG signals: (a) healthy subjects, (b) epileptic subjects during a seizure-free interval (interictal EEG), and (c) epileptic subjects during a seizure (ictal EEG). Each EEG is decomposed into the five constituent EEG sub-bands: delta, theta, alpha, beta, and gamma using wavelet-based filters. To identify the deterministic chaos in the system, the non-linear dynamics of the original EEGs are quantified in the form of the correlation dimension (CD, representing system complexity) and the largest Lyapunov exponent (LLE, representing system chaoticity). Similar to the original EEG, each sub-band is also subjected to chaos analysis to investigate the isolation of the changes in CD and LLE to specific sub-bands of the EEG. Subsequently, the effectiveness of CD and LLE in differentiating between the three groups is investigated based on statistical significance of the differences.

In order to extract the individual EEG sub-bands a wavelet filter is employed instead of the traditional Fourier transform because of the reasons described in Chapter 2. Similar to the signal decomposition described in Chapter 6, after a single level decomposition, two sequences are obtained representing the high and low resolution components of the signal, respectively. The low resolution components are further decomposed into low and high resolution components after a second level decomposition and so on. This multi-resolution analysis using four levels of decomposition yields five separate EEG sub-bands, which are subjected to subsequent chaos analysis. The process is explained in detail in the next section.

Following the procedure outlined in Chapter 3, the optimum lag and the minimum embedding dimension of the EEG are identified. Using these two parameters, the lagged phase space of the EEG can be constructed. These steps form the basis for the *preliminary chaos analysis* stage. The lagged phase space is important for identifying the temporal evolution of neuronal

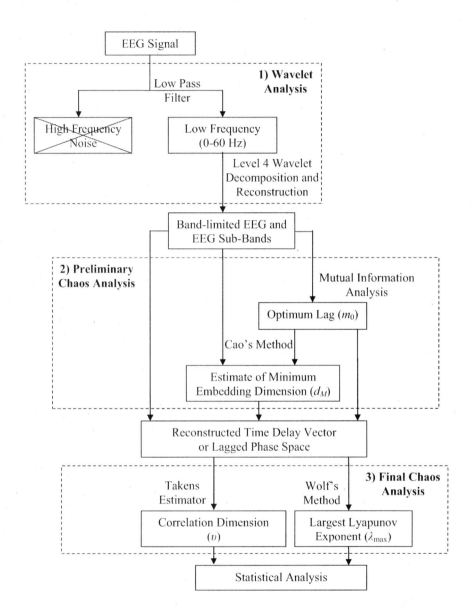

FIGURE 7.1
Wavelet-chaos algorithm for analysis of non-linear dynamics of EEGs

firing as represented by the EEG signal. In the *final chaos analysis* stage of the proposed wavelet-chaos algorithm, the CD is computed using the Takens estimator and the LLE is computed using a modification of Wolf's method reported in Iasemidis et al. (2000a).

7.3 Application and Results

7.3.1 Description of the EEG Data Used in the Research

The data used in this research are a subset of the EEG data for both healthy and epileptic subjects made available online by Dr. Ralph Andrzejak of the Epilepsy Center at the University of Bonn, Germany (`http://www.meb.uni-bonn.de/epileptologie/science/physik/eegdata.html`). EEGs from three different groups are analyzed: group H (healthy subjects), group E (epileptic subjects during a seizure-free interval), and group S (epileptic subjects during seizure). The type of epilepsy was diagnosed as temporal lobe epilepsy with the epileptogenic focus being the hippocampal formation. Each group contains 100 single channel EEG segments of 23.6 sec duration each sampled at 173.61 Hz (Andrzejak et al., 2001). As such, each data segment contains 4097 data points collected at intervals of 1/173.61 of a second. Each EEG segment is considered as a separate EEG signal resulting in a total of 300 EEG signals or EEGs.

Example EEGs for groups H (H029), E (E037), and S (S001) are displayed in Fig. 7.2. For the sake of visual clarity, the first six seconds of the three unfiltered EEGs are magnified in Fig. 7.3 (this is what a neurologist would normally read). The nomenclature in parentheses refers to the identifier num-

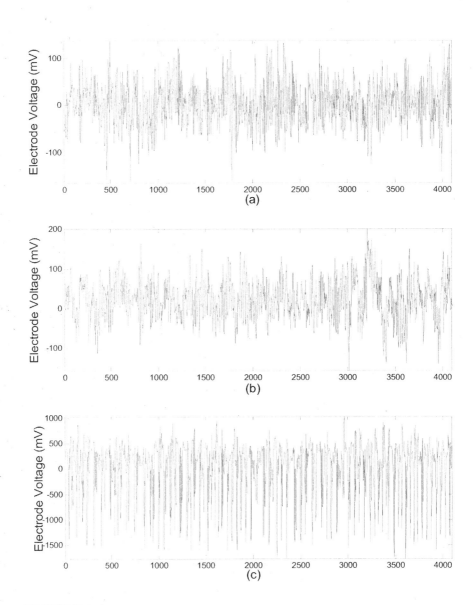

FIGURE 7.2
Unfiltered EEGs for (a) group H: healthy subject (H029), (b) group E: epileptic subject during a seizure-free interval (E037), and (c) group S: epileptic subject during seizure (S001)

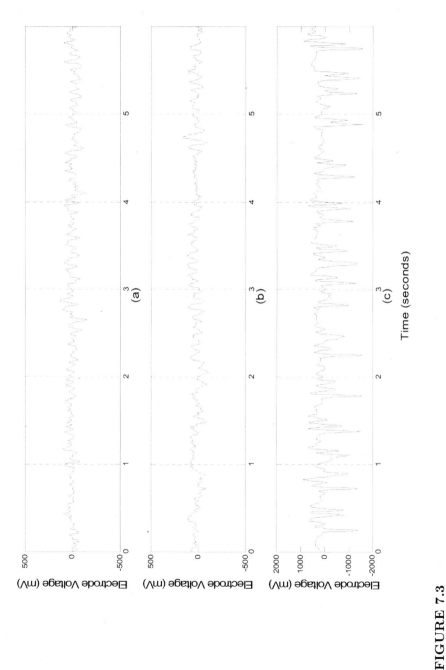

FIGURE 7.3

Unfiltered EEGs (0-6 seconds) for (a) group H: healthy subject (H029), (b) group E: epileptic subject during a seizure-free interval (E037), and (c) group S: epileptic subject during seizure (S001)

ber (group and signal number) for the EEG signal which will be used in the rest of the chapter.

7.3.2 Data Preprocessing and Wavelet Decomposition of EEG into Sub-Bands

The five primary EEG sub-bands, *delta*, *theta*, *alpha*, *beta*, and *gamma*, span the 0-60 Hz frequency range and higher frequencies are often characterized as noise. Since the sampling frequency of the EEG is 173.61 Hz, according to the Nyquist sampling theorem (Section 2.1), the maximum useful frequency is half of the sampling frequency or 86.81 Hz. As such, from a physiological standpoint, frequencies greater than 60 Hz can be classified as noise and discarded. Moreover, unlike the Fourier transform, wavelet decomposition does not allow the extraction of specific frequency bands without additional processing. Consequently, to correlate the wavelet decomposition with the frequency ranges of the physiological sub-bands, the wavelet filter used in this research (to be described shortly) requires the frequency content to be limited to the 0-60 Hz band. Due to the above mentioned reasons, the EEG is band-limited to the desired 0-60 Hz range by convolving with a low-pass finite impulse response (FIR) filter. The energy of the frequency band eliminated by the filter is negligible compared with that of the retained band in the range 0-60 Hz.

The band-limited EEG is then subjected to a level 4 decomposition using fourth order Daubechies wavelet transform. After the first level of decomposition, the EEG signal, **s** (0-60 Hz), is decomposed into its higher resolution components, **d1** (30-60 Hz) and lower resolution components, **a1** (0-30 Hz). In the second level of decomposition, the **a1** component is further decomposed into higher resolution components, **d2** (15-30 Hz), and lower resolution components, **a2** (0-15 Hz). Following this process, after four levels of decomposition, the components retained are **a4** (0-4 Hz), **d4** (4-8 Hz), **d3** (8-15

Hz), **d2** (15-30 Hz), and **d1** (30-60 Hz). Reconstructions of these five components using the inverse wavelet transform approximately correspond to the five physiological EEG sub-bands *delta, theta, alpha, beta,* and *gamma* (Fig. 7.4). Minor differences in the boundaries between the components compared to those between the EEG sub-bands are of little consequence due to the physiologically approximate nature of the sub-bands. Each EEG and its five sub-bands are subsequently subjected first to a preliminary and then to a final chaos analysis stage.

7.3.3 Results of Chaos Analysis for a Sample Set of Unfiltered EEGs

In order to compute the optimum lag, m_0, each EEG is tested for overlapping mutual information. The analysis involves computing the mutual information coefficients according to Eq. (3.4). The number of observations (N) is 4097 which is the number of data points in each EEG signal. When the number of probability states, N_S, is too small most data points are categorized as similar resulting in high mutual information coefficients. On the other hand, when N_S is too large the analysis becomes meaningless as all the data points are categorized as dissimilar and mutual information is completely lost. The appropriate value of this number is selected, for computational efficiency, to be a power of 2 and is determined by trial and error to be 128.

The optimum lag is at the first local minimum in the plot of the mutual information coefficient versus the lag time. The values of the optimum lag m_0 for the three EEGs (H029, E037, and S001) were found to be 5, 7, and 4, respectively. The minimum embedding dimensions for the EEGs are computed using Cao's method. The values of the modified embedding function $E1(d)$ (defined in Eq. 3.9) and $E2(d)$ (defined in Eq. 3.10) for the three EEGs (H029, E037, and S001) are shown in Fig. 7.5. The values of $E1(d)$ in all cases in Fig.

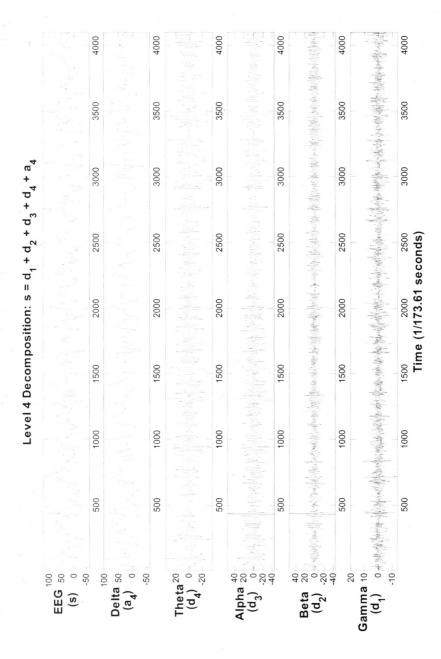

FIGURE 7.4

Level 4 decomposition of the band-limited EEG into five EEG sub-bands using fourth order Daubechies wavelet

7.5 approach 1 indicating chaos in the three sample EEGs obtained from both healthy and epileptic subjects. These figures also show that not all values of $E2(d)$ are equal to 1 and therefore the example EEGs all contain deterministic chaos. Convergence is assumed to be reached when the variation of the values of $E1(d)$ in three consecutive steps is within 5% of the maximum of all values. The minimum embedding dimensions, d_M, for the three EEGs (H029, E037, and S001) are obtained to be equal to 7.

The values of CD are obtained using the Takens estimator according to Eq. (3.15). The radius (ϵ) is varied from 0 to 20% of the maximum attractor size, i.e., the size of the lagged phase space. The maximum attractor size can be defined as the radius of a measuring unit that is just large enough to capture all the data points in the lagged phase space. The measuring unit is a circle for an embedding dimension of 2 and a sphere for an embedding dimension of 3. In the case of a very small radius not enough points are captured for the computation, whereas in the case of a large radius most of the available points are captured. Both these situations lead to an incorrect selection of nearest neighbors and yield inaccurate estimates of the CD. By trial and error, it is observed that the method yields fairly consistent estimates of CD within the range of 6-10% of maximum attractor size with highly variable estimates beyond that range. Therefore, the maximum value of the radius is set to 10% of the maximum attractor size. The values of CD for the three EEGs (H029, E037, and S001) are 7.0, 6.8, and 5.5, respectively. The values of the CD are less than or equal to the values for the minimum embedding dimension for all three sample EEGs, which is consistent with the theory that the minimum embedding dimension should be greater than the CD for any chaotic attractor (Natarajan et al., 2004).

The values of LLE are computed using the modification of Wolf's method

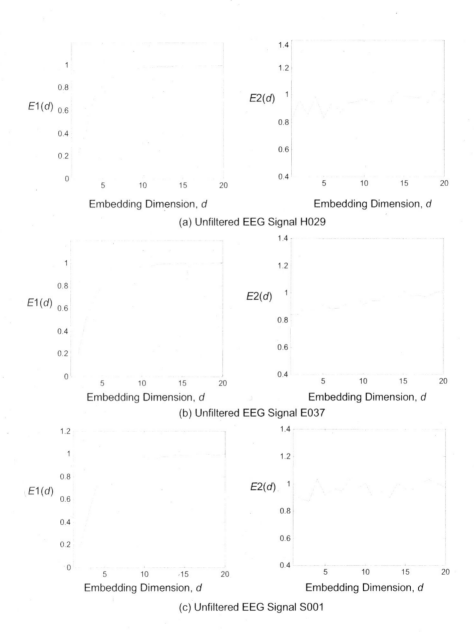

(a) Unfiltered EEG Signal H029

(b) Unfiltered EEG Signal E037

(c) Unfiltered EEG Signal S001

FIGURE 7.5

Plots of $E1(d)$ and $E2(d)$ versus embedding dimension for three sample EEGs:
(a) H029, (b) E037, and (c) S001

as explained in Iasemidis et al. (2000a). The values of LLE for the three EEGs (H029, E037, and S001) are 0.074, 0.037, and 0.067, respectively.

7.3.4   Statistical Analysis of Results for All EEGs

The chaos analysis presented in the previous section is repeated for all 300 band-limited EEGs (100 from each group H, E, and S) and for all five sub-bands of each EEG. This is a rather large EEG data set that can yield statistically significant results. The average values and standard deviations of the two parameters, CD and LLE, are computed for the band-limited EEGs and their sub-bands and summarized in Table 7.1 and Table 7.2, respectively. It is observed from these results that neither parameter by itself yields a sufficient method for quantification of the differences in the three groups. In some cases the range of values of CD and LLE of one group overlaps the values of another group. This indicates that a simple threshold applied to these parameters is insufficient to distinguish between the groups. Even so, it is clear that some of these reported variations are significant. Also, it should be noted that filtering of the signal alters the parameters employed to find the embedded attractor which may lead to a very different phase space behavior. Consequently, the properties of the phase space and the attractor for the sub-bands are no longer comparable directly to those for the original band-limited EEG.

From the CD values obtained from the band-limited EEGs (Table 7.1), it is observed that group S (5.3) differs from the other two groups, H (6.9) and E (6.7). However, the CD values for groups H and E do not appear to be significantly different from each other. The low value of CD for group S suggests a lowering of the complexity of the chaotic attractor during a seizure. Examination of the *delta* and *theta* sub-bands yields very similar values of CD for all three groups, H, E, and S (Table 7.1). The CD values in case of the *alpha* sub-band for group E (4.7) differs from the other two

TABLE 7.1
The average values of CD (with standard deviations in parenthesis) for band-limited EEGs and their five sub-bands in (a) group H: healthy subjects (100 signals), (b) group E: epileptic subjects during a seizure-free interval (100 signals), and (c) group S: epileptic subjects during seizure (100 signals)

Signal	Group H N=100	Group E N=100	Group S N=100
Band-limited EEG (0-60 Hz)	6.9 (1.4)	6.7 (1.2)	5.3 (1.3)
Delta (0-4 Hz)	5.9 (1.2)	5.7 (1.3)	5.4 (1.4)
Theta (4-8 Hz)	4.0 (0.5)	4.3 (0.8)	4.2 (0.6)
Alpha (8-12 Hz)	4.0 (0.5)	4.7 (1.0)	4.2 (0.8)
Beta (12-30 Hz)	4.5 (0.6)	4.1 (1.1)	3.5 (1.1)
Gamma (30-60 Hz)	3.7 (0.5)	3.1 (1.0)	2.6 (1.0)

TABLE 7.2
The average values of LLE (with standard deviations in parenthesis) for band-limited EEGs and their five sub-bands in (a) group H: healthy subjects (100 signals), (b) group E: epileptic subjects during a seizure-free interval (100 signals), and (c) group S: epileptic subjects during seizure (100 signals)

Signal	Group H N=100	Group E N=100	Group S N=100
Band-limited EEG (0-60 Hz)	0.089 (0.026)	0.041 (0.015)	0.070 (0.028)
Delta (0-4 Hz)	0.034 (0.007)	0.037 (0.008)	0.043 (0.012)
Theta (4-8 Hz)	0.096 (0.020)	0.082 (0.017)	0.080 (0.016)
Alpha (8-12 Hz)	0.106 (0.019)	0.078 (0.017)	0.086 (0.023)
Beta (12-30 Hz)	0.157 (0.037)	0.159 (0.031)	0.154 (0.033)
Gamma (30-60 Hz)	0.221 (0.064)	0.231 (0.073)	0.226 (0.066)

groups, H (4.0) and S (4.2) which do not appear to differ significantly from each other. This implies that if the *alpha* sub-band is considered to be a representation of brain dynamics by itself, then the attractor of the *alpha* sub-band, dubbed *alpha attractor*, has high complexity in epileptic patients during seizure-free intervals. The CD values for the *beta* and *gamma* sub-bands show considerable difference between all three groups. It is observed (Table 7.1) that both the *beta* and *gamma attractors* have the lowest complexity during a seizure, highest complexity for a healthy subject, and intermediate complexity for an epileptic subject during a seizure-free interval.

From the LLE values obtained from the band-limited EEGs (Table 7.2), it is observed that all three groups, H, E, and S, appear to differ from each other. The values suggest that the chaoticity of the chaotic attractor is highest in a healthy subject, lowest in an epileptic subject during a seizure-free state, and intermediate during a seizure. The LLE values for the *alpha* sub-band show considerable difference between all three groups and the same trend as in the case of the band-limited EEG. The LLE values in the case of the *theta* sub-band for group H (0.096) differs from the other two groups, E (0.082) and S (0.080), which do not appear to differ significantly from each other. This implies that the *theta* attractor has high chaoticity in healthy subjects. Examination of the *delta*, *beta*, and *gamma* sub-bands yields very similar values of LLE for all three groups, H, E, and S (Table 7.2).

In the ensuing discussions, one-way analysis of variance (ANOVA) is used for all statistical significance analysis performed at the 99% confidence level. Additionally, the significance of group differences is assessed using the method of Tukey's pairwise differences and the results are summarized in Table 7.3 along with p-values. Identical results are obtained using 99% confidence intervals (CI) ($\alpha = 0.01$) for both ANOVA and Tukey's method of pairwise differences. These results are consistent with the results discussed thus far in

TABLE 7.3

The results of the statistical analysis summarizing the groups that are separable using CD and LLE applied to the band-limited EEGs and the five physiological sub-bands. Consistent results are obtained using 99% confidence intervals ($\alpha = 0.01$) for both ANOVA and Tukey's method of pairwise differences. The p-values from the ANOVA are included in parenthesis.

Signal	Groups Differentiated By	
	CD	**LLE**
Band-limited EEG (0-60 Hz)	S (from E and H) ($p < 0.001$)	H, E, S ($p < 0.001$)
Delta (0-4 Hz)	-	S (from E and H) ($p < 0.001$)
Theta (4-8 Hz)	-	H (from E and S) ($p < 0.001$)
Alpha (8-12 Hz)	E (from H and S) ($p < 0.001$)	H, E, S ($p < 0.001$)
Beta (12-30 Hz)	H, E, S ($p < 0.001$)	-
Gamma (30-60 Hz)	H, E, S ($p < 0.001$)	-

this section. The CD differentiates between all three groups when computed based on the higher frequency *beta* and *gamma* sub-bands which have an identical pattern of high and low CD values (Fig. 7.6). Also, the statistical analysis reveals that the CD of the *alpha* sub-band distinguishes group E from both groups H and S, an observation not evident from Table 7.1 and Table 7.2. However, the CD for the *alpha* sub-band shows higher complexity for group E which is different from that noted in *beta* and *gamma* sub-bands (Table 7.3). This can be the reason why examination of the values of CD for the band-limited EEG differentiates only group S (from both groups E and H) but does not distinguish between groups E and H (Table 7.3). The CD of the lower frequency *delta* and *theta* sub-bands yields no discernable information at all (Table 7.3).

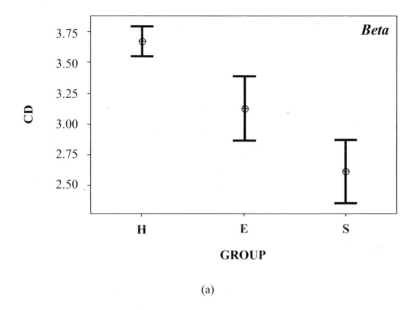

(a)

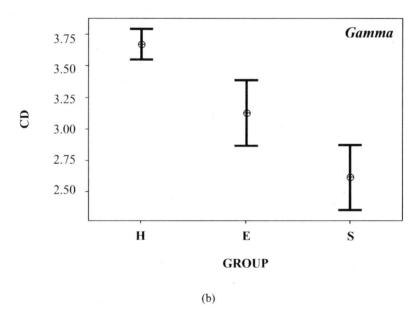

(b)

FIGURE 7.6
Confidence interval plots of CD values for (a) *beta* and (b) *gamma* sub-bands
showing significant differences between all three EEG groups H, E, and S

Unlike the CD, the LLE differentiates between all three groups when computed based on the intermediate frequency *alpha* sub-band. In the case of the lower frequency sub-bands, the *delta* LLE distinguishes group S (from both groups E and H) and *theta* LLE distinguishes group H (from both groups E and S) (Table 7.3). However, as recorded in Table 7.2, the patterns of high and low LLE values for these sub-bands are such that the effects do not contradict each other. Consequently, the LLE of the entire band-limited EEG also distinguishes between all three groups (Fig. 7.7). The LLE of the higher frequency *beta* and *gamma* sub-bands yields no discernable information at all.

7.4   Concluding Remarks

It is important to investigate the possible sources of differences in chaos parameters. Since the EEG is an overall representation of brain dynamics, it opens up the possibility that the observed changes in the parameters quantifying chaos in the band-limited EEG are actually the result of the superimposition of multiple processes underlying the EEG. One method of studying these underlying processes is to study the component physiological sub-bands of the EEG which can be assumed to represent these processes at a finer level.

The wavelet-chaos methodology presented in this chapter can be used to analyze the EEGs and *delta, theta, alpha, beta,* and *gamma* sub-bands of EEGs for detection of seizure and epilepsy. The non-linear dynamics of band-limited EEGs are quantified in terms of the CD and the LLE of the attractor to identify the deterministic chaos in the system.

The efficacy of CD and LLE as seizure and epilepsy detection parameters is investigated using a large number of EEGs. Statistical analysis of the values of these parameters shows that the LLE of healthy subjects, epileptic subjects

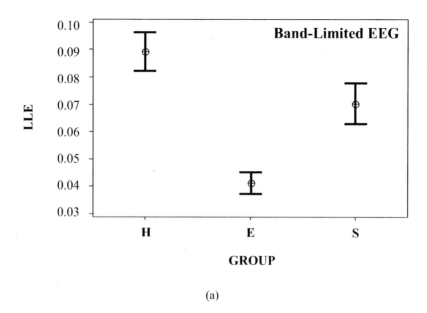

(a)

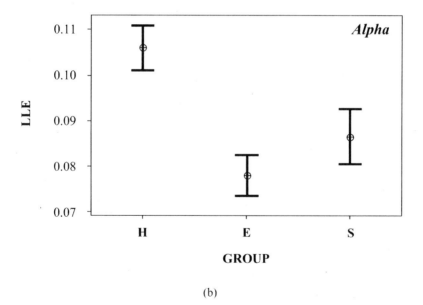

(b)

FIGURE 7.7
Confidence interval plots of LLE values for (a) band-limited EEG and (b) *alpha* sub-band showing significant differences between all three EEG groups H, E, and S

during a seizure-free interval, and epileptic subjects during seizure are significantly different from each other. The CD, on the other hand, is only capable of discriminating epileptic subjects during seizures from the other two groups. Although it is observed that, statistically, the LLE of the band-limited EEG can potentially distinguish between the three groups of subjects, it cannot be concluded with certainty that it will. If, based on a large number of EEG segments, the average values of the LLE or CD for the three groups are distinctly different, then better seizure detection can be expected but not guaranteed. The discovery of multiple potential discriminating parameters can result in increased accuracy for an effective real-time EEG epilepsy and seizure detection system.

To investigate the changes in CD and LLE of the physiological EEG sub-bands, a level 4 wavelet decomposition is performed to extract the *delta, theta, alpha, beta,* and *gamma* sub-bands. The original EEG is decomposed into five constituent sub-bands, each representing a subset of the processes underlying the overall brain dynamics. This decomposition process alters the original phase space and leads to new phase spaces that do not necessarily correspond directly to that of the original EEG. In other words, each sub-band is assumed to have its own chaotic attractor. Based on the statistical analysis of the sub-bands it is concluded that changes in the dynamics are not spread out equally across the spectrum of the EEG, but instead are limited to certain frequency bands. The lower frequency sub-bands (*delta, theta,* and *alpha*) show significant differences in terms of the LLE whereas the higher frequency sub-bands (*beta* and *gamma*) show significant differences in terms of the CD.

When the statistical analysis is based on the entire band-limited EEG, one may conclude that only the LLE (and not the CD) may be used as a discriminating parameter between the three groups. However, when the statistical analysis is performed on the EEG sub-bands, it is observed that the CD used within certain physiological sub-bands may also be employed to distinguish between all three groups. The availability of multiple potential discriminating parameters may result in increased accuracy of real-time EEG epilepsy and seizure detection systems, and is discussed in the following chapters.

8

Mixed-Band Wavelet-Chaos Neural Network Methodology

8.1 Introduction

Effective algorithms for automatic seizure detection and prediction can have a far-reaching impact on diagnosis and treatment of epilepsy. However, primarily due to a relatively low understanding of the mechanisms underlying the problem, most existing methods suffer from the drawback of low accuracy which leads to higher false alarms and missed detections (Iasemidis, 2003). Moreover, to obtain a reliable estimate of the efficacy of the epilepsy detection parameters and algorithms, they should be tested on a relatively large number of datasets. Due to the lack of reliable standardized data, most of the EEG analysis reported in the literature is performed on a small number of datasets which reduces the statistical significance of the conclusions. Such algorithms often demonstrate good accuracy for selected EEG segments but are not robust enough to adjust to EEG variations commonly encountered in a hospital setting.

In order to improve the statistical significance, a large number of EEG datasets belonging to three subject groups are used to investigate the performance of the methodology: a) healthy subjects (normal EEG), b) epileptic subjects during a seizure-free interval (*interictal* EEG), and c) epileptic sub-jects during a seizure (*ictal* EEG). The approach presented in the book is based on the premise that EEG sub-bands may yield more accurate infor-mation about constituent neuronal activities underlying the EEG. This was investigated in Chapter 7. Until recently, very little research was conducted to investigate the efficacy of combinations of measures selected from various phys-iological EEG sub-bands for use as classification parameters. As explained in Chapter 5, the problem of improving the classification accuracy is approached from two different angles: 1) designing an appropriate feature space by identi-fying combinations of parameters that increase the inter-class separation and 2) designing a classifier that can accurately model the classification problem based on the selected feature space.

8.2   Wavelet-Chaos Analysis: EEG Sub-Bands and Fea-ture Space Design

In Chapter 7, a wavelet-chaos methodology was presented for the analysis of EEGs and EEG sub-bands to identify potential parameters for seizure and epilepsy detection (Adeli et al., 2007). It was observed that when the values of the CD and LLE are computed from specific EEG sub-bands, the resulting differences in parameter values among the three groups are statistically signif-icant (at the 99% confidence level). Although no attempt was made to classify EEGs or detect seizure, based on the statistical conclusions it was theorized that a feature space comprising certain parameters computed from the EEG and its five sub-bands can enhance the accuracy of abnormality classification. In this work, in addition to CD and LLE, the standard deviation (STD) of the EEGs and EEG sub-bands is selected to characterize the signal variance.

Although STD, by itself, cannot differentiate all three groups, it is expected to increase the classification accuracy in combination with CD and LLE.

Prior to the wavelet decomposition, the input space consists of K EEGs from each of the three aforementioned groups (a total of $3K$ EEGs). Each EEG is represented as a time series vector, $\mathbf{X} = \{x_1, x_2, \ldots, x_N\}$ composed of a series of N single voltage readings at specific instants of time (indicated in the subscript). This nomenclature is identical to that used for the generic time series in Chapter 2. Following the wavelet decomposition, each EEG is decomposed into its five physiological sub-bands resulting in a total of six signals (including the band-limited EEG). This increases the size of the input space to $3K \times 6 (= 18K)$ signals (EEG and EEG sub-bands), each containing N data points. After the wavelet-chaos analysis, the feature space for each signal consists of three parameters: STD, CD, and LLE. The STD, CD, and LLE values obtained from the band-limited EEG as well as its five sub-bands yield a total of $6 \times 3 = 18$ parameters.

The parameters STD, CD, and LLE are represented by the vectors (of size $3K$) $\mathbf{S_D}$, $\mathbf{C_D}$, and $\mathbf{L_{LE}}$, respectively. Various combinations of these parameters are investigated as feature spaces for input to the classifiers with the goal of finding the most effective combination of parameters as well as the most effective classifier. Thus, the dimension of the input space is reduced from $N \times 18K$ to $P \times 3K$, where $P \in \{1, 2, \ldots, 18\}$ is the number of parameters used. Each EEG is then represented as a P-dimensional data point in the reduced input space, $\mathbf{F}$, which is dubbed *feature space*. Therefore, the purpose of the wavelet-chaos analysis is twofold. First, based on chaotic nonlinear dynamics of the brain it yields parameters with significant differences across the three groups of EEGs. This plays the role of feature enhancement and improves the subsequent classification accuracy. Second, it reduces a huge input space into a more manageable feature space.

In this chapter, four types of classifiers are investigated for classifying the EEGs into the aforementioned three groups: a) unsupervised k-means clustering, b) statistical discriminant analysis, c) radial basis function (RBF) neural network (RBFNN), and d) Levenberg-Marquardt backpropagation (BP) neural network (LMBPNN). Three different variations of the discriminant analysis are compared: linear discriminant analysis (LDA) using the Euclidean distance (ELDA) and the Mahalanobis distance (MLDA), and quadratic discriminant analysis (QDA). All three methods, ELDA, MLDA, and QDA, are explored because of the lack of knowledge about a) the underlying distribution and b) the required number of training datasets to achieve accurate classification results. Based on these studies, an innovative multi-paradigm wavelet-chaos-neural network methodology is presented for accurate classification of the EEGs. An overview of the methodology for the three-class EEG classification problem is presented schematically in Fig. 8.1.

8.3   Data Analysis

EEGs from the three groups are analyzed: group H (normal EEG), group E (interictal EEG), and group S (ictal EEG). Each group contains $K = 100$ single channel EEG segments of 23.6 sec duration, each sampled at 173.61 Hz (Andrzejak et al., 2001). As such, each data segment contains $N = 4097$ data points collected at intervals of $1/173.61$ of a second. Each EEG segment is considered as a separate EEG signal resulting in a total of 300 EEG signals or EEGs.

Testing the accuracy of all four classifiers with all different combinations of the eighteen parameters (2^{18} possible combinations) is a daunting task requiring months of computing time on a workstation. To reduce the computing

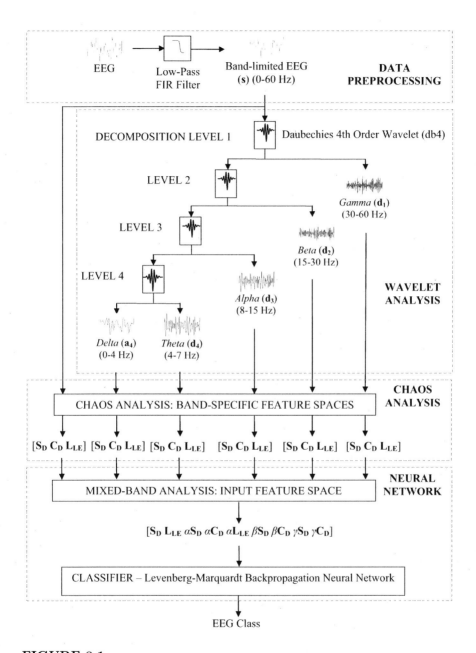

FIGURE 8.1

Overview of the wavelet-chaos-neural network methodology for the three-class EEG classification problem

time and output analysis to a more manageable one, the research is performed in two phases: a) band-specific analysis and b) mixed-band analysis. In the first phase, the six types of signals (EEG and the five EEG sub-bands) are considered one by one. For each type of signal, the EEGs are classified based on STD, CD, and LLE. Consequently, each signal is represented by a point in a 1- to 3-dimensional feature space. This is dubbed *band-specific analysis* in this research and the corresponding feature space is referred to as the *band-specific feature space*. One objective of band-specific analysis is to identify the classifiers that yield accurate classification and to eliminate the less accurate ones. Another objective is to identify specific combinations of parameters that may increase the classification accuracy of the selected classifiers.

In the second phase, the research is continued with the more promising classifiers and combinations of eighteen parameters selected from the six types of EEG and EEG sub-bands in the first step. As a result, each EEG is represented by a point in a 2- to 18-dimensional feature space. This is dubbed *mixed-band analysis* in this research and the corresponding feature space is referred to as the *mixed-band feature space*.

8.4   Band-Specific Analysis: Selecting Classifiers and Feature Spaces

8.4.1   *k*-Means Clustering

The centroid of a cluster representing any of the three EEG groups is initially selected randomly from the same group. The step is repeated for the remaining two groups. Beyond this *biased* initialization step, no prior information about group assignments of data points is utilized. The Euclidean distance is used

TABLE 8.1
Average classification accuracy percentages using k-means clustering for various band-specific feature spaces computed from the band-limited EEG and EEG sub-bands ($N_R = 100$; standard deviations in parentheses)

Signal	S_D	C_D	L_{LE}	$[S_D\,C_D]$	$[S_D\,L_{LE}]$	$[C_D\,L_{LE}]$	$[S_D\,C_D\,L_{LE}]$
				Parameter Combination			
Band-limited EEG	48.3 (0.0)	48.0 (0.7)	**59.3 (0.1)**	48.3 (0.0)	48.3 (0.0)	48.1 (0.9)	48.3 (0.0)
Delta (0-4 Hz)	52.3 (0.0)	40.3 (0.0)	45.9 (0.3)	52.3 (0.0)	52.3 (0.0)	40.3 (0.0)	52.3 (0.0)
Theta (4-7 Hz)	50.0 (0.0)	39.0 (0.4)	42.7 (0.6)	50.0 (0.0)	50.0 (0.0)	39.0 (0.4)	50.0 (0.0)
Alpha (8-12 Hz)	48.3 (0.0)	48.7 (0.5)	48.5 (0.3)	48.3 (0.0)	48.3 (0.0)	48.7 (0.5)	48.3 (0.0)
Beta (13-30 Hz)	49.3 (0.0)	44.0 (0.0)	36.3 (0.6)	49.3 (0.0)	49.3 (0.0)	44.0 (0.0)	49.3 (0.0)
Gamma (30-60 Hz)	48.0 (0.0)	49.2 (0.3)	37.1 (1.3)	48.0 (0.0)	48.0 (0.0)	48.9 (0.2)	48.0 (0.0)

as the proximity metric for the clustering process due to its simplicity. Once the clustering is complete, the assignment of the data points is compared with their known assignments to compute the classification accuracy. The sensitivity of k-means clustering to changes in starting points often leads to incorrect conclusions about the clustering accuracy. To overcome this shortcoming, the clustering process is repeated $N_R = 100$ times, with new points from each group selected randomly as initial points for every repetition. The average classification accuracy percentages obtained in 100 repetitions are tabulated in Table 8.1 (standard deviations are noted in parentheses). A maximum clustering accuracy of 59.3% is observed using LLE obtained from the band-limited EEG (identified with boldface in Table 8.1).

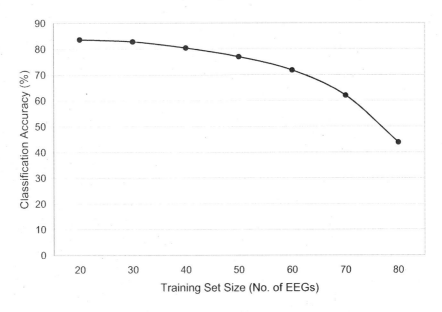

FIGURE 8.2
Classification accuracy of QDA using the feature space [$\mathbf{S_D}$ $\mathbf{C_D}$ $\mathbf{L_{LE}}$] for various training set sizes

8.4.2 Discriminant Analysis

In supervised classifications, for a given number of EEGs available, the dataset has to be divided carefully between training and testing sets. If the training set selected is too large, then the testing set becomes too small to yield meaningful classification accuracy results. In this research, considering 100 EEGs are available for each group, the size of the training dataset is varied from 20 to 50 EEGs (in increments of 10). Maximum classification accuracy was found when the training size is 20 EEGs for all three classifiers (ELDA, MLDA, and QDA) and input feature spaces. It was also found that the classification accuracy decreases with an increase in the size of the training dataset. As an example, the classification accuracy of QDA using the feature space [$\mathbf{S_D}$ $\mathbf{C_D}$ $\mathbf{L_{LE}}$] is plotted for training sizes of 20 to 80 EEGs in Fig. 8.2. The results for all three classifiers and other feature spaces display a similar trend.

To account for the effect of the training data on classification accuracy, the classification is repeated $N_R = 100$ times, each time with a new randomly selected training dataset. The average classification accuracy percentages are tabulated in Table 8.2 (standard deviations are noted in parentheses). The accuracies of three types of classifiers based on discriminant analysis are compared using various input feature spaces. Results obtained from the classifiers trained by a dataset of 20 EEGs only are presented in this chapter (Table 8.2) because this training size yields the highest classification accuracy, as mentioned previously. It is observed that, in general, the classification accuracy of QDA is the highest, followed by MLDA and then ELDA, with a few exceptions. The exceptions are for the combinations that yield low accuracies and consequently are of no interest in the rest of this research.

It is observed that feature spaces that are based on certain combinations of parameters computed from the EEG or specific EEG sub-bands yield more accurate results. In general, the feature spaces that show a higher classification accuracy are $[\mathbf{S_D}\ \mathbf{C_D}]$ when computed from the *beta* and *gamma* sub-bands, $[\mathbf{S_D}\ \mathbf{L_{LE}}]$ when computed from the band-limited EEG, and $[\mathbf{S_D}\ \mathbf{C_D}\ \mathbf{L_{LE}}]$ when computed from the *alpha* sub-band. These values are bold-faced in Table 8.2. Much lower classification accuracies are obtained from the *delta* and *theta* sub-bands. The two highest values of the classification accuracy percentages are obtained with QDA when the $[\mathbf{S_D}\ \mathbf{C_D}]$ feature space is computed from the *gamma* sub-band (85.5%) and the $[\mathbf{S_D}\ \mathbf{L_{LE}}]$ feature space is computed from the band-limited EEG (84.8%). These results validate the assertions made in Chapter 7 (Adeli et al., 2007) that the CD computed from the higher frequency *alpha*, *beta*, and *gamma* sub-bands and the LLE computed from the band-limited EEG and *alpha* sub-band can be instrumental in differentiating between the three groups.

TABLE 8.2
Average classification accuracy percentages using a training dataset size of 20 EEGs and discriminant analysis (ELDA, MLDA, and QDA) for various band-specific feature spaces computed from the band-limited EEG, and the *delta* and *theta* sub-bands ($N_R = 100$; standard deviations in parentheses)

Signal	Classifier	S_D	C_D	L_{LE}	$[S_D\ C_D]$	$[S_D\ L_{LE}]$	$[C_D\ L_{LE}]$	$[S_D\ C_D\ L_{LE}]$

<table>

Signal	Classifier	S_D	C_D	L_{LE}	$[S_D\ C_D]$	$[S_D\ L_{LE}]$	$[C_D\ L_{LE}]$	$[S_D\ C_D\ L_{LE}]$
Band-limited EEG	ELDA	47.3 (2.3)	36.1 (1.3)	52.9 (2.3)	47.6 (2.0)	79.6 (2.3)	68.6 (1.9)	81.1 (2.1)
	MLDA	53.1 (2.4)	35.3 (2.0)	49.8 (4.1)	54.9 (2.8)	84.0 (2.2)	66.1 (2.6)	81.6 (2.7)
	QDA	57.8 (3.1)	45.6 (1.4)	52.6 (2.7)	56.9 (3.0)	**84.8** **(3.1)**	66.4 (2.6)	83.5 (2.7)
Delta (0-4 Hz)	ELDA	60.4 (2.0)	21.6 (2.9)	31.0 (2.7)	53.4 (3.5)	57.1 (3.2)	29.4 (1.9)	54.0 (3.4)
	MLDA	58.9 (4.5)	21.0 (3.0)	28.8 (3.0)	55.5 (4.1)	60.1 (3.3)	27.8 (2.6)	57.4 (3.4)
	QDA	62.3 (2.7)	24.5 (2.8)	30.3 (2.8)	60.5 (3.4)	64.1 (2.9)	29.1 (2.1)	63.9 (2.7)
Theta (4-7 Hz)	ELDA	53.6 (2.8)	21.8 (4.3)	31.0 (3.7)	53.8 (3.7)	59.1 (2.9)	34.4 (3.6)	56.6 (2.7)
	MLDA	57.6 (2.2)	22.3 (2.8)	30.9 (3.7)	57.8 (3.0)	62.0 (4.2)	31.8 (3.7)	61.0 (4.0)
	QDA	60.5 (2.6)	23.9 (2.2)	27.8 (3.5)	65.3 (3.3)	63.0 (4.0)	33.3 (3.2)	65.1 (3.0)
Alpha (8-12 Hz)	ELDA	60.1 (1.9)	34.1 (2.1)	33.4 (1.4)	67.3 (1.8)	66.4 (2.2)	38.0 (1.9)	68.4 (2.1)
	MLDA	63.4 (3.5)	32.3 (2.8)	31.1 (1.9)	64.0 (4.4)	73.1 (2.8)	38.4 (3.2)	69.9 (3.3)
	QDA	65.4 (2.4)	38.3 (2.0)	35.4 (2.0)	74.5 (2.9)	75.4 (1.8)	42.6 (2.4)	**75.9** **(2.4)**
Beta (13-30 Hz)	ELDA	72.3 (2.0)	30.1 (2.1)	16.4 (1.9)	58.6 (7.1)	56.8 (7.7)	30.9 (2.0)	55.3 (5.2)
	MLDA	74.6 (2.1)	27.9 (3.0)	17.3 (2.5)	73.0 (3.6)	69.5 (4.4)	31.0 (2.7)	72.0 (4.5)
	QDA	75.4 (1.6)	31.8 (2.0)	17.8 (2.4)	**80.9** **(2.8)**	74.6 (2.1)	33.3 (1.6)	80.1 (2.9)
Gamma (30-60 Hz)	ELDA	71.8 (2.7)	35.6 (1.0)	17.1 (2.9)	59.0 (4.1)	51.9 (9.8)	35.1 (2.4)	55.6 (4.6)
	MLDA	74.5 (2.9)	32.0 (3.0)	15.9 (4.1)	71.1 (5.0)	68.8 (5.4)	29.6 (4.5)	68.4 (5.6)
	QDA	81.0 (1.7)	36.6 (1.7)	17.1 (3.2)	**85.5** **(1.2)**	78.8 (2.6)	37.6 (2.4)	83.6 (1.9)

Parameter Combination (Training Size = 20 EEGs)

8.4.3 RBFNN

Two parameters of RBFNN architecture and formulation are investigated: 1) the number of nodes in the hidden layer and 2) the spread of the RBF, p, with the goal of achieving optimum network performance. The number of nodes in the hidden layer affects the training performance. Employing a large number of nodes, for example, equal to the number of training instances is computationally expensive but enables the network to be trained with 100% accuracy. Training accuracy is defined as $(1 - \text{training error}) \times 100\%$. In many cases, employing a smaller number of nodes yields sufficiently accurate results. Therefore, training is started with a small number of nodes and is repeated with an increasingly larger number of nodes. Training is terminated when any one of two conditions holds: 1) training error decreases to 0.001 or 2) the number of nodes is equal to the number of training instances.

The spread, p, has to be selected very carefully. In RBFNN, a specific input is supposed to excite only a limited number of nodes in the hidden layer. When the spread is too large, all hidden nodes respond to a given input which results in loss of classification accuracy. On the other hand, when the spread is too small, each node responds only to a very specific input. Consequently, the node will be unable to classify any new input accurately. The optimum spread for maximum classification accuracy is found by varying it from 0.5 to 5 in increments of 0.5. The training and classification accuracies versus the spread values are presented in Fig. 8.3. It is observed that for the most part the classification training accuracy decreases whereas the classification (testing) accuracy increases as the spread is increased. Both curves reach approximately a plateau for larger values of the spread, say, equal to or greater than 4. The reciprocal relationship may be attributed to the overtraining of the RBFNN and its consequent inability to respond to new inputs appropriately.

The classification training-testing process is repeated $N_R = 10$ times, each

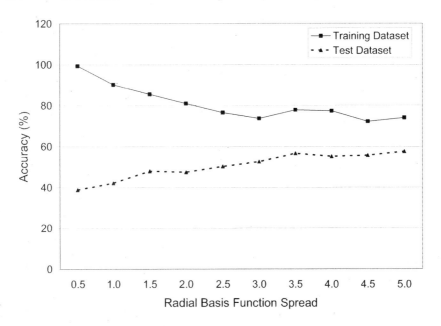

FIGURE 8.3
Variations of the RBFNN training and testing classification accuracies versus
the RBF spread

time with a new randomly selected training dataset. Since, typically, neural
network training requires significantly larger computational efforts compared
with k-means clustering or discriminant analysis, the number of repetitions is
reduced to ten. The average classification accuracy percentages for all band-
specific feature spaces using a training size of 40 EEGs and an RBF spread of 4
are tabulated in Table 8.3 (standard deviations are noted in parentheses). It is
observed that RBFNN requires a larger training size than the classifiers based
on discriminant analysis but yields less accurate results. Also, the standard
deviations of the results are much larger for RBFNN than for discriminant
analysis which implies greater dependence of RBFNN on the training data.
The two highest values of classification accuracy are obtained when the $[\mathbf{S_D}]$
feature space is computed from the *gamma* sub-band using a spread of 4 and
training sizes of 40 (76.5%) (shown by boldface in Table 8.3) and 50 (76.2%)

TABLE 8.3
Average classification accuracy percentages using a training dataset size of 40 EEGs and RBFNN (RBF spread = 4) for various band-specific feature spaces computed from the band-limited EEG and EEG sub-bands ($N_R = 100$; standard deviations in parentheses)

Signal	Parameter Combination (Training Size = 40 EEGs, RBF Spread = 4)						
	S_D	C_D	L_{LE}	$[S_D C_D]$	$[S_D L_{LE}]$	$[C_D L_{LE}]$	$[S_D C_D L_{LE}]$
Band-limited EEG	41.9 (12.3)	43.4 (3.1)	44.3 (2.9)	47.0 (11.0)	52.1 (6.7)	61.9 (2.7)	51.9 (11.8)
Delta (0-4 Hz)	53.8 (7.7)	38.9 (1.8)	40.8 (1.6)	41.0 (7.7)	47.7 (7.4)	41.3 (3.6)	48.1 (4.7)
Theta (4-7 Hz)	50.2 (11.5)	37.1 (1.8)	34.1 (0.9)	45.8 (6.8)	45.8 (9.8)	42.6 (2.8)	51.0 (4.1)
Alpha (8-12 Hz)	49.7 (9.5)	40.5 (3.2)	33.6 (1.6)	52.9 (5.7)	38.2 (6.9)	42.2 (2.5)	58.8 (7.9)
Beta (13-30 Hz)	58.5 (6.7)	37.1 (3.4)	33.1 (1.0)	59.1 (6.8)	49.5 (9.0)	39.8 (3.4)	64.8 (6.1)
Gamma (30-60 Hz)	**76.5** **(3.0)**	39.5 (4.6)	34.1 (0.8)	55.3 (4.9)	66.8 (5.8)	42.2 (3.4)	51.2 (9.0)

(not shown in Table 8.3). Results obtained with other spread values in the range of 0.5 to 5 are not shown in this book for the sake of brevity.

8.4.4 LMBPNN

LMBPNN is investigated in a manner similar to RBFNN in terms of the training dataset sizes and the number of classification training-testing repetitions. Overall, a training size of 40 EEGs yields the best classification accuracy and is therefore selected for all EEGs and EEG sub-bands. To accurately model the complex dynamics underlying EEGs, the effect of the network architecture, that is, the number of hidden layers and the number of nodes in the hidden layer on the classification accuracy is investigated. This was performed in a manner different from that of RBFNN due to basic differences in the architecture of the two neural networks. The number of nodes in the first and second hidden layers is increased in increments of 5 from 5 to 20 and 0 to 15, respectively. Training is terminated when any one of three conditions holds: 1) training error decreases to 0.001, 2) training error gradient decreases to 0.01, or 3) number of training epochs reaches 100. Based on this parametric study two hidden layers each with 10 to 15 nodes appeared to yield the best classification results for all combinations of the parameters computed from the six types of EEGs and EEG sub-bands. No other discriminatory patterns were observed.

The average classification accuracy percentages for all band-specific feature spaces for a training size of 40 are summarized in Table 8.4 (standard deviations are noted in parentheses). In general, it is observed that the feature spaces that show higher classification accuracy are $[\mathbf{S_D}\ \mathbf{C_D}]$ when computed from the *beta* and *gamma* sub-bands and $[\mathbf{S_D}\ \mathbf{L_{LE}}]$ when computed from the band-limited EEG and the *alpha* sub-band. These values are identified by boldface in Table 8.4. Much lower classification accuracy is obtained from the *delta* and *theta* sub-bands. This observation is similar to that observed for the case of QDA. The two highest values of classification accuracy are obtained when the $[\mathbf{S_D}\ \mathbf{L_{LE}}]$ feature space is computed from the band-limited EEG (89.9%) and the $[\mathbf{S_D}\ \mathbf{C_D}]$ feature space is computed from the *gamma* sub-band (87.3%).

8.5   Mixed-Band Analysis: Wavelet-Chaos-Neural Network

Figure 8.4 summarizes the maximum classification accuracy percentages of all band-specific feature spaces for six different types of classifiers obtained in phase one of this research. The results of the extensive band-specific analysis

TABLE 8.4

Average classification accuracy percentages using a training dataset size of 40 EEGs and LMBPNN for various band-specific feature spaces computed from the band-limited EEG and EEG sub-bands ($N_R = 100$; standard deviations in parentheses)

Signal	Parameter Combination (Training Size = 40 EEGs)						
	S_D	C_D	L_{LE}	$[S_D\,C_D]$	$[S_D\,L_{LE}]$	$[C_D\,L_{LE}]$	$[S_D\,C_D\,L_{LE}]$
Band-limited EEG	64.8 (2.7)	47.2 (3.3)	56.7 (6.8)	63.8 (3.2)	**89.9** **(1.8)**	72.1 (3.4)	86.9 (2.9)
Delta (0-4 Hz)	68.2 (8.5)	39.9 (4.2)	41.1 (2.1)	63.9 (4.0)	66.1 (6.3)	40.2 (3.3)	65.1 (8.0)
Theta (4-7 Hz)	69.4 (4.5)	39.2 (2.5)	48.2 (4.6)	72.8 (4.4)	71.6 (4.5)	43.8 (3.9)	73.8 (4.1)
Alpha (8-12 Hz)	74.3 (4.3)	42.8 (2.9)	42.1 (3.9)	79.9 (3.3)	**80.5** **(2.7)**	48.7 (5.0)	79.4 (6.7)
Beta (13-30 Hz)	81.1 (2.0)	43.1 (5.1)	38.1 (3.9)	**85.5** **(2.1)**	79.7 (2.7)	45.2 (4.6)	83.4 (1.8)
Gamma (30-60 Hz)	86.4 (2.0)	44.9 (3.8)	46.9 (3.4)	**87.3** **(1.7)**	83.1 (2.6)	50.0 (4.1)	85.7 (2.7)

lead to the conclusion that QDA and LMBPNN yield the two highest classification accuracies (85.5% and 89.9%, respectively). Moreover, both classifiers have small standard deviations using various combinations of training/testing datasets. Consequently, they are robust with respect to changes in training data. In general, LMBPNN yields a higher classification accuracy than QDA, but requires a larger training dataset size (40 EEGs versus 20 EEGs in the parametric studies performed in this research). In phase two of the research, QDA and LMBPNN are selected for further investigation using the 2- to 18-dimensional mixed-band feature spaces described earlier.

Based on the band-specific analysis of phase one, two other important conclusions are made. For LMBPNN, the following eight parameters yield higher classification accuracies: STD computed from the band-limited EEG and *alpha*, *beta*, and *gamma* sub-bands; CD computed from *beta* and *gamma* sub-bands; and LLE computed from the band-limited EEG and *alpha* sub-

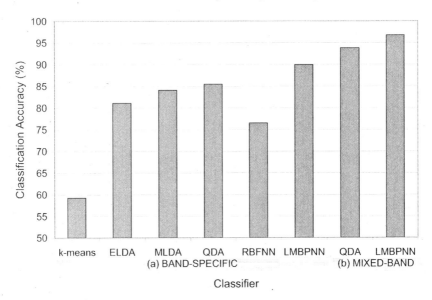

FIGURE 8.4
Maximum classification accuracy percentages obtained from a) six different types of classifiers for all band-specific feature spaces and b) QDA and LMBPNN for the mixed-band feature space [$\mathbf{S_D}$ $\mathbf{L_{LE}}$ $\alpha\mathbf{S_D}$ $\alpha\mathbf{C_D}$ $\alpha\mathbf{L_{LE}}$ $\beta\mathbf{S_D}$ $\beta\mathbf{C_D}$ $\gamma\mathbf{S_D}$ $\gamma\mathbf{C_D}$].

band. The same parameters yield higher classification accuracies for QDA as well, plus a ninth parameter, that is, CD computed from the *alpha* sub-band.

In the second phase, over five hundred different combinations of mixed-band feature spaces consisting of promising parameters from the first phase were investigated. This effort took weeks of workstation computing time. For statistical consistency, the same number of training-testing repetitions of $N_R = 10$ was used for both QDA and LMBPNN. A training size of 40 EEGs yielded the highest classification accuracies for the mixed-band feature spaces and was employed for both classifiers.

It was discovered that a mixed-band feature space consisting of all nine aforementioned parameters [$\mathbf{S_D}$ $\mathbf{L_{LE}}$ $\alpha\mathbf{S_D}$ $\alpha\mathbf{C_D}$ $\alpha\mathbf{L_{LE}}$ $\beta\mathbf{S_D}$ $\beta\mathbf{C_D}$ $\gamma\mathbf{S_D}$ $\gamma\mathbf{C_D}$] results in the highest classification accuracy for both QDA and LMBPNN. In this notation, the parameter prefix denotes the EEG sub-band from which

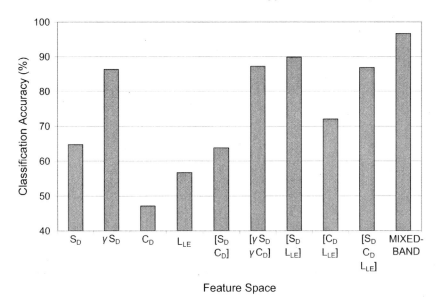

FIGURE 8.5

Comparison of maximum classification accuracy percentages of LMBPNN for different feature spaces. The mixed-band feature space is $[\mathbf{S_D} \ \mathbf{L_{LE}} \ \alpha\mathbf{S_D} \ \alpha\mathbf{C_D} \ \alpha\mathbf{L_{LE}} \ \beta\mathbf{S_D} \ \beta\mathbf{C_D} \ \gamma\mathbf{S_D} \ \gamma\mathbf{C_D}]$.

the parameter is computed. Absence of a prefix indicates that the parameter is computed from the band-limited EEG. The classification accuracies obtained using the nine-parameter mixed-band feature space for both QDA and LMBPNN are also presented in Fig. 8.4.

The average classification accuracy percentages using QDA and LMBPNN are 93.8% (with a standard deviation of 1.0) and 96.7% (with a standard deviation of 2.9), respectively. The classification accuracy of LMBPNN using the nine-parameter mixed-band feature space is significantly improved compared with using various band-specific feature spaces, as shown in the sample results of Fig. 8.5. Band-specific feature spaces based on single parameters yield low to very low classification accuracies (<65.0%) except for the STD computed from the *gamma* sub-band which yields a classification accuracy of 86.4%. Band-specific feature spaces based on combinations of two and three parameters lead to more accurate classification (in the range of 65-90%). The nine-parameter mixed-band feature space yields the highest classification accuracy (96.7%). Such a high level of accuracy has not been reported previously in the literature.

The training of the LMBPNN classifier with the nine-parameter mixed-band feature space presented in the chapter is completed rather quickly. On an Intel Pentium IV workstation (1700 MHz, 512 MB RAM), the classifier training for ten repetitions is completed in approximately 30 seconds for 120 training instances (40 from each of the three subject groups). Using the trained network to classify a new nine-parameter mixed-band input takes only a fraction of a second on the aforementioned workstation.

8.6   Concluding Remarks

In this chapter, a novel wavelet-chaos-neural network methodology is presented for the analysis of EEGs for detection of seizure and epilepsy and applied to three different groups of EEG signals obtained from healthy and epileptic subjects. Four classification methods were investigated: k-means cluster analysis, discriminant analysis, RBFNN, and LMBPNN. A disadvantage of an unsupervised classification method such as k-means clustering is that addition of new data changes the cluster centroids and, therefore, the cluster assignments. k-Means clustering yielded the lowest classification accuracy compared to other methods indicating low intra-cluster similarities and/or low inter-cluster dissimilarities.

Discriminant analysis, in general, yields much higher classification accuracy than k-means clustering. QDA yields the best classification accuracy followed by MLDA and then ELDA. The disadvantages of QDA, mentioned earlier, do not appear to have much of an impact on the classification process. As compared to LDA, no significant increase in computational time or training size was observed. QDA yielded higher classification accuracies for datasets of the same size as LDA, and decreasing the training size did not alter this trend. RBFNN training converges rapidly and yields a completely trained network (100% training accuracy). It yields lower classification accuracies than all three types of discriminant analysis as well as LMBPNN. LMBPNN yields the highest value of classification accuracy among all classifiers both for band-specific analysis (89.9%) as well as for mixed-band analysis (96.7%). A similar improvement over band-specific analysis is also observed when mixed-band analysis is employed with QDA.

Based on these analyses using the wavelet-chaos methodology, the nine-parameter mixed-band feature space $[\mathbf{S_D}\ \mathbf{L_{LE}}\ \alpha\mathbf{S_D}\ \alpha\mathbf{C_D}\ \alpha\mathbf{L_{LE}}\ \beta\mathbf{S_D}\ \beta\mathbf{C_D}\ \gamma\mathbf{S_D}\ \gamma\mathbf{C_D}]$ yields the most accurate classification results. When this feature space is used as input to the LMBPNN classifier, a maximum classification accuracy of 96.7% is achieved.

All three key components of the wavelet-chaos-neural network methodology are important for improving the accuracy of EEG classification. Wavelet analysis decomposes the EEG into sub-bands and is instrumental in creating the mixed-band feature spaces with an improved accuracy over band-specific feature spaces. The parameters used in these feature spaces are obtained by statistical analysis (STD) and chaos analysis (CD and LLE). Although these parameters are unable to classify the EEGs individually, their combinations improve the classification accuracy significantly especially when multiple-parameter mixed-band feature spaces are employed. Finally, the LMBPNN classifier classifies the EEGs into the three groups more accurately than any of the other classification methods investigated. This illustrates the point that a judicious combination of parameters and classifiers can accurately discriminate between the three types of EEGs, which in turn could lead to increased accuracy of real-time epilepsy and seizure detection systems.

9

Principal Component Analysis-Enhanced

Cosine Radial Basis Function Neural

Network

9.1 Introduction

In a clinical setting, a vast number of physiological variables contribute to and affect the EEG. As a result, the distinctions between the different EEG groups are not very well defined. For the computer model to be clinically effective, it is required to be robust with respect to EEG variations across subjects and various mental states. The ultimate objective is a comprehensive tool for epilepsy diagnosis as well as real-time monitoring of EEGs for seizure detection and, eventually, seizure prediction. Toward this end, the following classifier characteristics are desirable: fast training, high classification accuracy for all three groups, and low sensitivity to training data and classifier parameters, that is, robustness, to be discussed later in this chapter. Although a high classification accuracy is reported using LMBPNN in Chapter 8, the class of RBFNN classifiers is investigated further, in this chapter, to increase the robustness of classification.

The original feature space consists of K EEGs from each of the three aforementioned groups (a total of $3K$ EEGs) where each EEG is composed of N voltage readings at specific instants of time. The wavelet decomposition of each EEG results in a total of six signals (the band-limited EEG and its five physiological sub-bands). This increases the size of the feature space to $3K \times 6 = 18K$ signals (EEG and EEG sub-bands), each containing N data points. Following the wavelet-chaos analysis described in Chapter 3, each EEG is represented by nine parameters (or features) and therefore is reduced to a nine-dimensional data point in the mixed-band feature space. Thus, the dimensions of the feature space are reduced from $18K \times N$ to $3K \times 9$.

For supervised learning, the available input dataset denoted by a matrix $\mathbf{F}$ of dimension $3K \times 9$ is divided into training input ($\mathbf{F_R}$) and testing input ($\mathbf{F_T}$) sets. The training input consists of k training instances out of the $3K$ available EEGs where each instance is represented by the nine aforementioned features. The number of training instances, k, is called training size. An equal number of EEGs ($k \div 3$) is selected from each group for training the neural network. Therefore, the training and testing inputs are matrices of size $4k \times 9$ (dimensions of $\mathbf{F_R}$) and $(3K - k) \times 9$ (dimensions of $\mathbf{F_T}$), respectively. The actual training and testing classifier outputs are denoted by $\mathbf{C_R}$ with dimensions $k \times 1$ or $k \times 3$ and $\mathbf{C_T}$ with dimensions $(3K - k) \times 1$ or $(3K - k) \times 3$, respectively (the number of columns in each matrix depends on the output encoding to be described later). It should be noted that these matrices have been transposed from their generalized forms described in Chapter 3 to simplify matrix operations.

In this chapter, a new two-stage classifier is presented for accurate and robust EEG classification based on the nine-parameter mixed-band feature space discovered in Section 8.5. In the first stage, principal component analysis (PCA) is employed for feature enhancement by transforming the nine-parameter feature space into a new feature space which is more amenable to subsequent classification. PCA is usually used for dimensionality reduction to reduce the computational requirement, and in some cases, for denoising (Zhukov et al., 2000; Lee and Choi, 2003). Although dimensionality reduction is also achieved in this work, that is not the primary objective. Rather, the rearrangement of the input space along the principal components of the data is used to improve the classification accuracy of the cosine radial basis function neural network (RBFNN) employed in the second stage. The PCA-enhanced cosine RBFNN classifier proposed in this chapter not only is a novel methodology, but also the first such application in EEG classification. It will be shown that the integration of the mixed-band wavelet-chaos methodology (Ghosh-Dastidar et al., 2007) and the new PCA-enhanced cosine RBFNN yields high EEG classification accuracy and is quite robust with respect to changes in training data, a common concern in applications of supervised learning. An overview of the proposed EEG classification methodology is presented schematically in Fig. 9.1.

9.2   Principal Component Analysis for Feature Enhancement

The training input, $\mathbf{F_R}$, is subjected to PCA for the purpose of determining the principal components. At this stage, the testing input is kept completely separate so as not to contaminate the analysis. Following the steps detailed in Section 4.5, the following five steps are performed to determine the axes of the new reoriented input space (Smith, 2002).

1. The centroid of the k training instances, denoted by the 1×9 vector $\mu_\mathbf{R}$, is found.

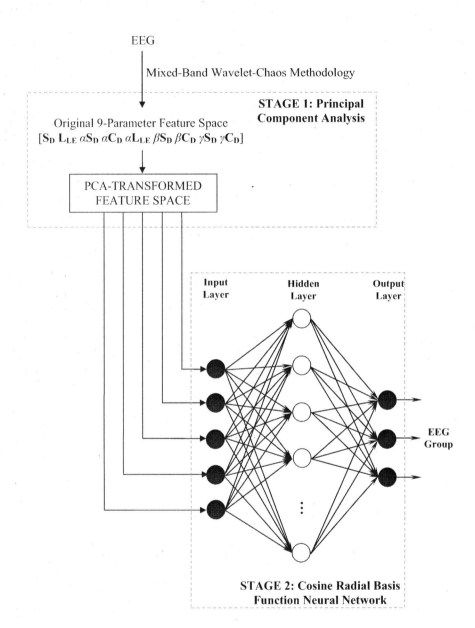

FIGURE 9.1

Overview of the wavelet-chaos-neural network methodology for the three-class EEG classification problem and architecture of the two-stage PCA-enhanced cosine RBFNN classifier with one hidden layer

2. The origin of the coordinate system is moved to the centroid of the training instances by subtracting $\mu_{\mathbf{R}}$ from each point in $\mathbf{F_R}$.

3. The pairwise covariance between all nine parameters constituting the columns of the shifted training input is computed. This is represented by a 9×9 covariance matrix, denoted by $cov(\mathbf{F_R} - \mu_{\mathbf{R}})$.

4. The eigenvectors and eigenvalues of the covariance matrix are computed resulting in nine 9×1 eigenvectors and the nine corresponding eigenvalues. The eigenvectors are mutually orthogonal and represent the axes of the new coordinate system. The eigenvector corresponding to the maximum eigenvalue is the principal component.

5. Eigenvectors corresponding to the lowest eigenvalues are discarded. The remaining n eigenvectors are arranged column-wise in the order of decreasing eigenvalues to form the $9 \times n$ eigenvector matrix, $\mathbf{E}$. How many eigenvectors to keep is found by trial and error, as described shortly in this chapter.

Following PCA, the training and testing inputs (consisting of nine features) are transformed into the new input space as follows:

$$\mathbf{F_{RX}} = \mathbf{F_R} \times \mathbf{E} \tag{9.1}$$

$$\mathbf{F_{TX}} = \mathbf{F_T} \times \mathbf{E} \tag{9.2}$$

The transformed training ($\mathbf{F_{RX}}$) and testing ($\mathbf{F_{TX}}$) input datasets consist of n features (corresponding to the selected eigenvectors) and, therefore, are matrices of size $k \times n$ and $(3K - k) \times n$, respectively. Since n is smaller than the original number of features (i.e., 9), $\mathbf{F_{RX}}$ and $\mathbf{F_{TX}}$ are smaller in size compared with $\mathbf{F_R}$ and $\mathbf{F_T}$, respectively.

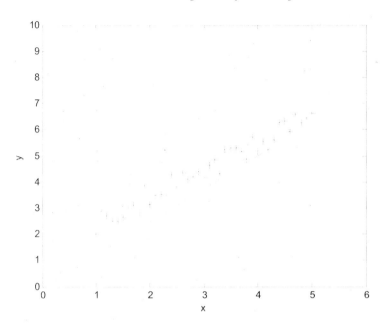

FIGURE 9.2
A simple two-dimensional example to demonstrate the effect of PCA on the
angular spread: data in the original coordinate system

The reorientation of the feature space is expected to increase the angular separation between individual data points and lead to improved cosine RBFNN classification accuracy. Since the EEG feature space is nine-dimensional and cannot be visualized, a simple two-dimensional example, shown in Figs. 9.2 and 9.3, is used to illustrate this concept graphically. The data plotted in Fig. 9.2 after transformation using PCA are shown in Fig. 9.3. It is observed in Fig. 9.3 that the transformed data has approximately 60% wider angular spread compared with the original data in Fig. 9.2 when viewed from the origins of the respective coordinate systems, and thus will be more amenable to accurate classification.

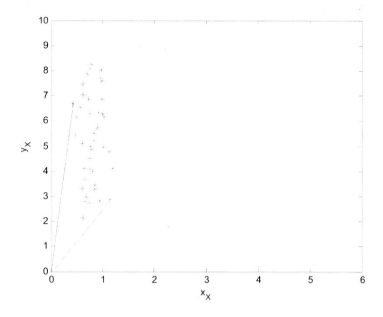

FIGURE 9.3
A simple two-dimensional example to demonstrate the effect of PCA on the angular spread: data transformed using PCA

9.3 Cosine Radial Basis Function Neural Network: EEG Classification

In the last decade, RBFNNs have been applied to problems of function approximation and pattern recognition in different fields with great success (Adeli and Karim, 2000; Howlett and Jain, 2001a,b; Karim and Adeli, 2002a, 2003; Adeli and Karim, 2005). Recently, neural network-based classifiers have also been employed in the nascent field of automated EEG analysis and epilepsy diagnosis (Lee and Choi, 2003; Güler et al., 2005; Mohamed et al., 2006; Ghosh-Dastidar et al., 2007). However, in spite of the architecture simplicity and guaranteed convergence of RBFNNs, these classifiers have not been investigated for seizure detection. Besides the incorporation of PCA, there are two

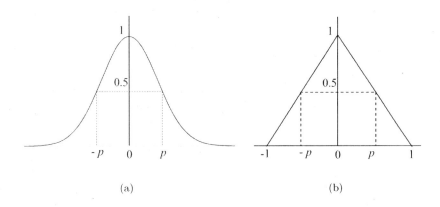

(a)   (b)

FIGURE 9.4
Shapes of the basis functions (a) RBF and (b) TBF

primary differences between the RBFNN presented in this chapter compared with a classical RBFNN. First, the commonly used Euclidean distance is replaced by an angular measure as input to the RBFNN hidden layer nodes. Second, the radial basis function (RBF) for hidden layer nodes is replaced by a triangular basis function (TBF). Although technically this changes the classifier to a *triangular* basis function neural network, the name *radial basis function neural network* is retained as it represents the entire class of neural networks. The shapes of the two basis functions, RBF and TBF, are shown in Fig. 9.4(a) and Fig. 9.4(b), respectively. As explained in the next section, the cosine angular distance and the TBF are selected because they yield higher classification accuracies as compared with the Euclidean distance and RBF.

The RBFNN classifier consists of one hidden layer in addition to the input and output layers. The input layer consists of n nodes corresponding to the n transformed features after PCA is performed. The hidden layer can have a maximum of k nodes, equal to the number of training instances. Employing such a large (k) number of nodes is computationally expensive but enables the network to be trained with no training convergence error. In many cases, employing a smaller number of nodes, $N_h < k$, yields sufficiently accurate results. Therefore, training is started with a small number of nodes and is repeated with an increasingly larger number of nodes until the desired training convergence is achieved. The number of nodes in the output layer, r, depends on the output encoding scheme to be discussed in the next section. The RBFNN architecture with three output nodes ($r = 3$) is shown in Fig. 9.1.

The weights of the links connecting the n input nodes to the N_h nodes in the hidden layer are represented by the $n \times N_h$ weight matrix $\mathbf{W}$. The weights of the links connecting the input layer to the jth hidden node are denoted by the $n \times 1$ column vector $\mathbf{W}(j)$. The $n \times 1$ input column vector for the mth training instance ($1 \leq m \leq k$) is denoted by $\mathbf{F_{RX}}(m)$. The weighted input to the jth node in the hidden layer for the mth training instance is computed as the angle between the two column vectors $\mathbf{F_{RX}}(m)$ and $\mathbf{W}(j)$ as:

$$\mathbf{I}(j) = \cos^{-1}\left(\frac{\mathbf{W}(j) \cdot \mathbf{F_{RX}}(m)}{|\mathbf{W}(j)|\,|\mathbf{F_{RX}}(m)|}\right) \tag{9.3}$$

where the numerator is the dot product of the two vectors and the denominator is the product of the magnitudes of the two vectors. The weighted input $\mathbf{I}$ is a column vector of size $N_h \times 1$. Using a triangular basis transfer function shown in Fig. 9.4(b), the output of the jth node of the hidden layer is computed as:

$$\mathbf{Y}(j) = \begin{cases} 1 - \left|\dfrac{\mathbf{I}(j)}{2p}\right| & \forall \mathbf{I}(j) \in (-2p, 2p) \\ 0 & \text{otherwise} \end{cases} \tag{9.4}$$

where the spread, p, scales the output such that the output is 0.5 (the average of the limits of 0 and 1) when the weighted input $\mathbf{I}(j)$ is equal to p. When $\mathbf{W}(j)$ is equal to $\mathbf{F_{RX}}(m)$, the weighted input is 0 which results in an output of 1.

The input to the output layer nodes is computed as the weighted sum of the outputs from the hidden layer nodes. The error function employed for training the neural network is the sum of squared errors (SSE), E, as follows:

$$E = \sum_{j=1}^{r} [\mathbf{C_R}(j) - \mathbf{O_R}(j)]^2 \qquad (9.5)$$

where $\mathbf{C_R}$ and $\mathbf{O_R}$ are the actual and desired classifier outputs, respectively, with dimensions $k \times r$.

9.4   Applications and Results

9.4.1   Neural Network Training

Each iteration of RBFNN training involves the addition of a hidden layer node j with the input weight vector, $\mathbf{W}(j)$, equal to the input vector, $\mathbf{F_{RX}}(m)$, for the training instance, m, that minimizes the error. Training is terminated when any one of two conditions holds: 1) SSE is reduced to a limiting convergence value determined by numerical experimentation to balance the training and testing errors and avoid overtraining (Ghosh-Dastidar et al., 2007) or 2) the number of nodes becomes equal to the number of training instances. A limiting convergence value of 0.05 was obtained for the EEG data used in this research.

9.4.2   Output Encoding Scheme

Two output encoding schemes are investigated in this work. In scheme 1, there is only a single output node yielding one of the three values, -1, 0, and 1, representing the three EEG groups of signals, respectively (Ghosh-Dastidar et al., 2007). The second scheme employs three output nodes corresponding to the three EEG groups. The node returns an output of 1 if the EEG belongs to the corresponding group and zero otherwise. Therefore, the three correct classifications are encoded as $\{1, 0, 0\}$, $\{0, 1, 0\}$, or $\{0, 0, 1\}$.

9.4.3 Comparison of Classifiers

In Chapter 8, the accuracy of various classifiers was investigated for the three-group EEG classification problem based on the nine-parameter mixed-band feature space. Six different classifiers were trained multiple times with a randomly selected training input dataset out of the 300 available EEGs and the average classification accuracies were computed. The Levenberg-Marquardt backpropagation neural network (LMBPNN) classifier yielded the highest average accuracy of 96.7% with a standard deviation of 2.9% representing the sensitivity to the training data. A classical RBFNN classifier (based on an RBF and the Euclidean distance) yielded low classification accuracies in the vicinity of 80% and a high standard deviation in the neighborhood of 8% but required less training time.

Based on these observations, it appeared that either 1) the choice of parameters defining the RBFNNs (such as the spread) was sub-optimal or 2) classical RBFNNs were unsuitable for the EEG classification problem. The first issue is addressed by performing an extensive parametric and sensitivity analysis to find the optimum RBFNN parameters for maximum classification accuracy and minimum standard deviation. To account for the effect of the training data on classification accuracy, the classification is repeated $N_R = 100$ times, each time with a new randomly selected training dataset. Next, a PCA-enhanced cosine RBFNN neural network is presented which models the EEG classification problem more accurately.

The new methodology is compared with seven other RBFNN classifiers created by varying three characteristics: basis function (RBF or TBF), distance function (Euclidean distance abbreviated as Euc and angular cosine distance abbreviated as Cos), and PCA (employed or not employed) as well as the LMBPNN classifier. The RBFNN classifiers are tagged based on the combination of the aforementioned characteristics as follows: 1) classical RBFNN (RBF+Euc), 2) RBF+Euc+PCA, 3) RBF+Cos, 4) RBF+Cos+PCA, 5) TBF+Euc, 6) TBF+Euc+PCA, 7) TBF+Cos, and 8) PCA-enhanced cosine RBFNN (TBF+Cos+PCA). The average classification accuracy percentages for all nine classifiers using two different output encoding schemes are tabulated in Table 9.1 (standard deviations are noted in parentheses). The highest classification accuracy and lowest standard deviation are obtained for the PCA-enhanced cosine RBFNN proposed in this chapter (noted in boldface in Table 9.1). The results of Table 9.1 show that, for the RBFNN classifiers (the first eight in Table 9.1), the second output encoding scheme yields consistently higher classification accuracies by 2.8-5.3% and lower standard deviations compared with the first scheme. It is noted that the LMBPNN classifier exhibits behavior different from that of the RBFNN classifiers. Scheme 2 for LMBPNN yields lower classification accuracies and higher standard deviations compared with scheme 1. Therefore, it appears that scheme 2 is better suited for the RBFNN classifiers only. Therefore, only the second output encoding scheme is included in the remainder of the chapter for the RBFNN classifiers.

Using the output encoding scheme 2, the first two classifiers RBF+Euc and RBF+Euc+PCA yield the lowest classification accuracies of 95.1%. If either the Euclidean distance is replaced with the cosine distance (classifier 3) or the RBF is replaced with TBF (classifier 5), the classification accuracy increases to 95.7%. When cosine distance and TBF are used (classifier 7) the classification accuracy increases to 96.2%. Addition of PCA (classifier 8) further increases the classification accuracy to 96.6%. The standard deviation values for eight classifiers are in the range 1.4-1.7%. The standard deviation

TABLE 9.1
The average classification accuracy percentages for eight classifiers (standard deviations are noted in parentheses)

No.	Classifier	Output Encoding Scheme 1	Output Encoding Scheme 2
1	RBF+Euc	92.1 (2.5)	95.1 (1.7)
2	RBF+Euc+PCA	92.1 (2.4)	95.1 (1.6)
3	RBF+Cos	90.9 (2.3)	95.7 (1.6)
4	RBF+Cos+PCA	90.5 (2.3)	95.8 (1.6)
5	TBF+Euc	92.8(1.8)	95.7 (1.6)
6	TBF+Euc+PCA	92.8 (2.0)	95.6 (1.6)
7	TBF+Cos	93.0 (2.1)	96.2 (1.5)
8	TBF+Cos+PCA	92.8 (2.1)	**96.6 (1.4)**
9	LMBPNN	96.7 (2.9)	89.9 (4.0)

for classifier 8 is the lowest (1.4%) which leads to the conclusion that classifier 8 is the most robust to changes in training data. The classification accuracy is comparable to the LMBPNN classifier but the sensitivity to the choice of training data is reduced to half.

It is observed that employing PCA with the Euclidean distance-based RBFNN does not improve the classification accuracy because the reorientation of the feature space does not affect the Euclidean distance. Employing TBF yields higher classification accuracies than using RBF. An extensive study involving weeks of computational time was performed in which, in addition to RBF and TBF, the following three activation functions were also investigated: Mexican hat wavelet, Morlet wavelet, and a truncated sinc (sinx/x) function. The spread of these functions was defined in a manner similar to TBF (Eq. 9.5) where, as explained earlier, an input equal to the spread resulted in an output of 0.5. However, employing these functions resulted in much lower classification accuracies (80-90%), and therefore they are not discussed in detail.

The first derivative of TBF is discontinuous at three points $(\mathbf{I}(j) = -1, 0,$

and 1 in Eq. 9.5). This affects the RBFNN node behavior in two ways. First, at $\mathbf{I}(j) \leq -1$ and $\mathbf{I}(j) \geq 1$, the TBF function value becomes zero without the smooth asymptotic transition observed in the other four activation functions. Therefore, values of $\mathbf{I}(j) \leq -1$ and $\mathbf{I}(j) \geq 1$ do not contribute at all to the output of the RBFNN node in the case of TBF whereas for the other functions there is some small contribution. Second, inputs within the range $\mathbf{I}(j) \in (-1, 1)$ have a much greater contribution to the output of the RBFNN node in TBF than in the case of the other functions. This functional behavior seems to fit the EEG data more accurately.

9.4.4   Sensitivity to Number of Eigenvectors

The input to the cosine RBFNN consists of n features corresponding to the selected eigenvectors. As a result, the classifier is required to have n input nodes. Generally speaking, a large value of n leads to a large number of input nodes which exponentially increases the complexity of the neural network classifier and the computational effort required. This is known as the *dimensionality curse* (Bellman, 1961). As such, it is desired that the value of n be as small as possible without compromising the classification accuracy. This is not a significant issue in the current research because n is limited to a maximum value of 9. The issue was studied nevertheless. The change in classification accuracy of PCA-enhanced cosine RBFNN with the number of eigenvectors is presented in Fig. 9.5. The classification accuracy increases with the number of eigenvectors and plateaus at $n = 5$ with a value of 96.6%. It should be noted that this does not imply that five out of the nine original features are sufficient for the classification problem. Rather, it means that five or more linear combinations of *all* nine features can model the classification problem accurately.

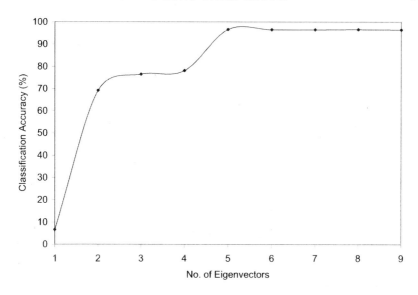

FIGURE 9.5
Variations in classification accuracy of PCA-enhanced cosine RBFNN with
the number of eigenvectors

9.4.5   Sensitivity to Training Size

For a given dataset a decision has to be made with regard to the sizes of the
training and testing data. The training size, k, affects the classification accuracy of the neural network and needs to be selected carefully. On the one hand,
if k is too small the classifier is unable to model the classification problem accurately. On the other hand, if k is too large, the size of the remaining data
to be used as a testing set would be too small to test the model effectively.
In this work, the sensitivity of the PCA-enhanced cosine RBFNN is assessed
by increasing the training size from $k = 60$ (20 EEGs from each group) to
$k = 240$ (80 EEGs from each group) in increments of 30 (10 for each group).
The change in classification accuracy of PCA-enhanced cosine RBFNN with
training size is shown in Fig. 9.6. The classification accuracy increases rapidly
at first, but plateaus near 96.6% for training sizes greater than 150 EEGs.
A training size of 240 EEGs yields a slightly higher classification accuracy of

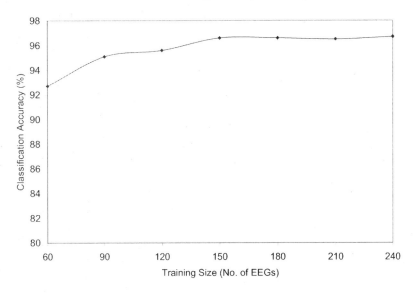

FIGURE 9.6
Variations in classification accuracy of PCA-enhanced cosine RBFNN with training size

96.7%. This value is not reported in Table 9.1 because using the large training size $k = 240$ EEGs results in a small testing dataset of only 60 EEGs. A training size of $k = 150$ (50 EEGs from each group) is deemed sufficient to model this problem. At the same time, the plateau in classification accuracy for larger training sizes indicates that the model stabilizes beyond a certain training size.

9.4.6 Sensitivity to Spread

The spread p of the RBF usually plays an important role in determining the classification accuracy of RBFNN (Ghosh-Dastidar et al., 2007). In such networks, a specific input is supposed to excite only a limited number of nodes in the hidden layer. When the spread is too large, all hidden nodes respond to a given input, which results in loss of classification accuracy. On the other hand, when the spread is too small, each node responds only to a

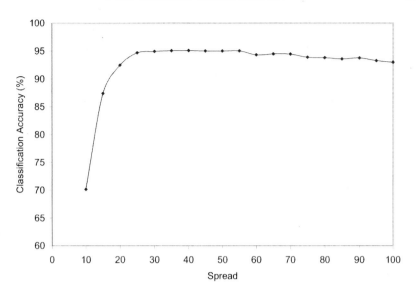

FIGURE 9.7
Classification accuracy versus spread for RBFNN

very specific input and therefore is unable to classify any new input accurately. The effect of varying the spread from 10 to 100 in increments of 5 on the classical RBFNN is illustrated in Fig. 9.7. It is observed that the classification accuracy is very low for $p = 10$ and increases rapidly until $p = 25$. The maximum classification accuracy (95.1%) is obtained in the range $p \in (25, 60)$ and then slowly decreases to 93.0% for $p = 100$. The classification accuracy continues to decrease for values of p larger than 100 (not shown in Fig. 9.7).

Figure 9.7 shows clearly the classification accuracy depends on the choice of the spread significantly. Two different approaches can be used to attack this problem: 1) developing an algorithm to compute the optimum value of p automatically for a given training input set and 2) designing a classifier that is less sensitive to the spread. An inherent disadvantage of the first approach is that the spread computation is dependent on the training data which makes it sensitive to data outliers. Moreover, this approach relies on retrospective data. In a prospective clinical application, the testing input is unknown and

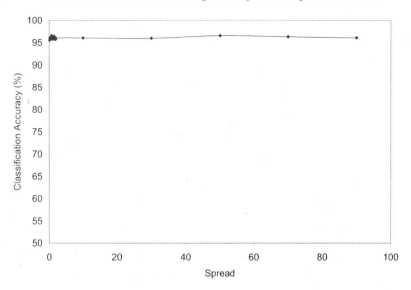

FIGURE 9.8
Classification accuracy versus spread for PCA-enhanced cosine RBFNN

unclassified and as a result, evaluation of the *correctness* of the spread by trial and error is not possible. No method currently exists to accurately estimate the spread. Therefore, the second approach is advanced in this research to increase the reliability of the classification.

The effect of varying the spread on the PCA-enhanced cosine RBFNN is illustrated in Fig. 9.8. There is very little variation in the classification accuracy (95.8-96.6%) for a wide range of spread values (0 to 90). The classifier was tested extensively with spread values in the range from 0 to 2 which is the reason for the dense population at the beginning of the graph in Fig. 9.8. The proposed PCA-enhanced cosine RBFNN is quite robust with respect to variations in the spread.

9.5   Concluding Remarks and Clinical Significance

A novel PCA-enhanced cosine radial basis function neural network classifier is presented. The new two-stage classifier is integrated with the mixed-band wavelet-chaos methodology, developed in Chapters 7 and 8, for accurate and robust classification of three different groups of EEG signals obtained from healthy and epileptic subjects (Ghosh-Dastidar et al., 2007).

In order to assess the clinical applicability of the proposed wavelet-chaos-neural network methodology for epilepsy diagnosis and seizure detection, the average classification accuracies for each group (H, E, or S) are tabulated in Table 9.2 (standard deviations are noted in parentheses). The values reported are obtained using the PCA-enhanced cosine RBFNN with the following characteristics: TBF spread $(p) = 1.0$, training size $(k) = 50$, output encoding scheme 2, and number of retained principal components $(n) = 6$. The training and testing of the classifier is repeated $N_R = 10$ times, each time with a new randomly selected training dataset.

It is observed that the group-wise classification accuracies for normal EEGs (group H), interictal EEGs (group E), and ictal EEGs (group S) are 98.4%, 97.0%, and 94.8%, respectively. A normal EEG is misclassified as an ictal EEG 0.6% of the time. The model is unable to classify a normal EEG into any of the three groups 1% of the time. A normal EEG is never misclassified as an ictal EEG and vice versa. An interictal EEG is misclassified as an ictal EEG 2.6% of the time and not classified at all 0.4% of the time. An ictal EEG is misclassified as a normal EEG 1.0% of the time, an interictal EEG 3.0% of the time, and not classified 1.2% of the time. From the perspective of seizure detection, the false alarm rate is 2.6+0.6=3.2% and the missed detection rate is 1.0+3.0+1.2=5.2%.

TABLE 9.2
Group-wise distribution of the average classification accuracy ($N_R = 10$) of the PCA-enhanced RBFNN classifier (standard deviations are noted in parentheses). A *percentage value* at the intersection of *row name* and *column name* in the table should be read as "The percentage of EEGs from group *row name* that were classified as group *column name* is *percentage value*". H, E, and S represent the three subject groups and X represents data not classified by the proposed model

	H	E	S	X
H	98.4 (2.3)	0.0 (0.0)	0.6 (1.3)	1.0 (1.7)
E	0.0 (0.0)	97.0 (3.3)	2.6 (3.1)	0.4 (0.8)
S	1.0 (1.1)	3.0 (1.4)	94.8 (2.3)	1.2 (1.7)

Practicing neurologists have the most difficulty in differentiating between interictal and normal EEGs. For epilepsy diagnosis, when only normal and interictal EEGs are considered, the classification accuracy of the proposed model is 99.3%. This statistic is especially remarkable because even the most highly trained neurologists do not appear to be able to detect interictal EEGs more than 80% of the time.

Part III

Automated EEG-Based Diagnosis of Alzheimer's Disease

10

Alzheimer's Disease and Models of Computation: Imaging, Classification, and Neural Models

---

10.1   Introduction

Old age is naturally associated with some loss of memory and cognition. The clinical condition in which these symptoms become severe enough to hamper social and occupational functions is referred to as dementia. Alzheimer's disease (AD), the leading cause of dementia, is a neurodegenerative disorder that affects about 5 million people in the United States (Khachaturian and Radebaugh, 1996). The majority of the victims are above the age of 65. This late-onset type of AD is classified as senile dementia of the Alzheimer type (SDAT) and it affects 7% of people above the age of 65 and 40% above 80 (Price, 2000). Early-onset cases of AD in people in their 40s and 50s are less common and are classified as presenile dementia of the Alzheimer type (PDAT). This chapter presents a review of research in imaging, classification, and neural models of AD. This review does not distinguish between PDAT and SDAT but refers to them collectively as AD. AD is considered to be a *complex* disorder (Mayeux, 1996). In other words, no single factor is necessary and sufficient to cause the disease. Multiple types of influences such as genetic, environmental, and dietary factors contribute to the progression of the disease but the causes and underlying mechanisms are not known with any measure of certainty.

Any kind of dementia, including AD, primarily causes memory impairment (long-term or short-term) and at least one of the following classes of clinical symptoms: impairment of cognition, impairment of judgment, change in personality, and altered behavior (American Psychiatric Association, 1994). The early symptoms of AD are usually mild but subsequently progress in severity with age. Attempts have been made to identify the transition from symptoms of normal aging to AD. Recently, a transition state, mild cognitive impairment (MCI), characterized by loss of short-term memory, has been identified as a likely precursor to AD (Shah et al., 2000; Morris et al., 2001; Saykin and Wishart, 2003). However, even though MCI increases the risk of developing AD, there is no conclusive evidence of a direct connection or clearly defined boundaries.

Diagnosis of AD poses a challenge even for trained neurologists for multiple reasons (Heyman, 1996; Price, 2000). First, the symptomatology of AD is similar to that resulting from other dementias associated with cerebrovascular diseases, Lewy body dementia, Parkinson's disease, depression, nutritional deficiencies, and drug reactions. Second, sometimes dementias may be due to more than one of the above mentioned causes which makes diagnosis even more complicated. Third, the lack of clearly defined boundaries between symptoms of normal aging, MCI, and the onset of AD can lead to diagnostic errors. Therefore, a person suffering from symptoms of probable AD is required to undergo a battery of clinical tests and a diagnosis is made based on patient history and elimination of all other possible causes. Currently, there is no single clinical test for diagnosis of AD even in its later stages, and most recent research is focused on developing a noninvasive, sensitive, and specific in vivo test. In fact, at the present, AD can only be confirmed from a post mortem autopsy of brain tissue from the presence of structures in the brain such as neurofibrillary tangles and senile plaques (Khachaturian and Radebaugh, 1996).

Prediction or early-stage diagnosis of AD would require a more comprehensive understanding of the underlying mechanisms of the disease and its progression. Researchers in this area have approached the problem from multiple directions by attempting to develop (a) neurological (neurobiological and neurochemical) models, (b) analytical models for anatomical and functional brain images, (c) analytical feature extraction models for electroencephalograms (EEGs), (d) classification models for positive identification of AD, and (e) neural models of memory and memory impairment in AD. The first four approaches are correlated in the sense that any neurological change will lead to physical changes that will be manifested in some form in brain images and EEGs which can then be detected using classification techniques. The neural models can provide an alternative method for understanding and explaining these changes. These changes are called diagnostic markers if they occur during the progression of AD or prediction markers if they occur before the advent of AD.

10.2 Neurological Markers of Alzheimer's Disease

AD is a neurodegenerative disease which, by definition, causes neuronal death and dysfunction. It has been hypothesized that neuronal death occurs as a result of loss of neuronal synapses (Terry, 1996). This results in both anatomical (structural) as well as physiological (functional) changes or markers in various regions of the brain. Postmortem examination of brain tissue reveals that the frontal, temporal, and parietal lobes of the brain are affected the most. Since AD is a progressive disease, the magnitude of these markers depends on the stage of the disease. Further, the spatial and temporal distribution of these markers is not uniform. For instance, the regions associated with memory are affected earlier than others. It should be noted that in a highly complex system such as the human brain, the dysfunction of a single region, type of cell, or neurotransmitter cannot be conclusively identified as the sole factor responsible for these changes. Damage to one area of the brain spreads to other areas via neuronal pathways in the brain resulting in the damage and dysfunction of multiple other areas which may lead to a wide range of symptoms. Therefore, it is very important that the entire system is treated as a whole and all aspects are examined carefully before arriving at any conclusions.

From a gross external examination of the affected regions of the brain of an Alzheimer's patient, changes in certain structures in the affected regions have been identified, in addition to the observations of generalized cortical atrophy characterized by reduced gyri volume and increased sulci width. Neuronal damage leads to atrophy of structures in the frontal, temporal, and parietal lobes. In accordance with the heterogeneous symptomatology of AD, the damage is localized to various sites within these lobes. These sites include the neocortex, entorhinal cortex, hippocampus, amygdala, nucleus basalis, anterior thalamus, corpus callosum, and brain stem nuclei such as the substantia nigra, locus ceruleus, and the raphe complex (Mirra and Markesbery, 1996; Xanthakos et al., 1996; Price, 2000; Teipel et al., 2003). Ventricles (especially the lateral ventricle) also appear to be enlarged which may be a primary symptom or a secondary effect associated with atrophy of the periventricular parenchyma. Another consistent feature seen in AD is the abnormal paleness of the locus ceruleus, a nucleus of neurons containing neuromelanin. In some cases, the substantia nigra also loses its characteristic black color (similar to that seen in Parkinsonism).

A microscopic examination of the affected regions yields neuropathological structures such as neurofibrillary tangles and senile plaques that have been commonly established to be markers of AD. Neurofibrillary tangles are left behind in the affected regions after neuron death caused by abnormalities in the neuronal cytoskeletal structure. The neuronal cytoskeleton is responsible for maintaining the cell structure as well as for transportation and exchange of molecules. In AD, an important microtubule-binding protein, tau, becomes hyperphosphorylated which disrupts the normal axonal transport mechanisms leading to impaired movement of various molecules and, eventually, neuron death (Price, 2000; Avila et al., 2002).

Senile plaques, the other common marker of AD, are classified into two sub-types - neuritic and diffuse plaques based on their structure (Mirra and Markesbery, 1996). Neuritic plaques have a well defined spherical structure with a periphery of neurites (axons or dendrites of damaged neurons, astrocytes, and microglia) surrounding a central dense core of amyloid protein. On the other hand, diffuse plaques have a less defined, amorphous structure and lack neurites. Diffuse plaques may be found in normal aging brain tissue to some extent and there is some debate as to whether diffuse plaques may be precursors of neuritic plaques or not (Selkoe, 1994; Mirra and Markesbery, 1996). Amyloid angiopathy, a related marker, involves the deposition of the amyloid protein in cerebral blood vessels in AD patients. This has also been detected to some extent in normal aging brains, but still may be important as a potential diagnostic marker (Tian et al., 2003; de Courten-Myers, 2004; Haglund et al., 2004).

The nature of the neuronal damage, in addition to being region specific, is also cell specific. In other words, the damage is restricted to specific populations of neurons and specific neurotransmitters in specific regions of the brain (Price, 2000). The primary cells affected in the neocortex and the entorhinal cortex are the glutaminergic pyramidal neurons and, to some extent, interneurons. The pyramidal cells are severely affected in the CA1 and CA2 regions of the hippocampus. Damage to these areas is a critical factor responsible for memory impairment, the primary symptom in AD. Another group of neurons affected in AD are the cholinergic neurons in the basal forebrain (primarily, the nucleus basalis, medial septal nucleus, and the diagonal band of Broca) which project to the neocortex and hippocampus (Mirra and Markesbery, 1996; Price, 2000). Damage to these neurons destroys the activity of the neurotransmitter acetylcholine in the destination regions resulting in memory and attention deficits. However, whether this is one of the causes of the disease or an effect is not known (Mesulam, 2004). Damage to the other sites such as the amygdala, thalamus, and the brain stem nuclei is usually responsible for disruption of behavior, emotions, and the associative aspects of memory.

Changes in cerebral blood flow, glucose metabolism, oxidative free radical damage to mitochondrial DNA, neuroreceptor functioning, and neurotransmitter activity have also been identified as potential markers for the study of AD (Budinger, 1996). Studies have demonstrated reduced cerebral blood flow or perfusion as a result of neuronal death (especially in the temporal and temporoparietal regions) in AD patients as compared to a control group of normal subjects (Reed et al., 1989; Pearlson et al., 1992; Bozzao et al., 2001). There is also evidence of reductions in glucose metabolism (Hoyer, 1996; Planel et al., 2004) and increased oxidative free radical damage (Evans, 1993; Eckert et al., 2003; Baloyannis et al., 2004) in the same regions. Due to abnormal neurotransmitter activity in the brains of AD patients, different types of neurotransmitter receptors show abnormal behavior. Studies have confirmed the decline in the density of nicotinic acetylcholine, serotonin, and

α_2-epinephrine receptors (Budinger, 1996) which reduces neurotransmitter binding and therefore synaptic efficiency. The deficit of acetylcholine and abnormality in the concentrations of other neurotransmitters such as serotonin and gamma-aminobutyric acid (GABA) that affect the synthesis and actions of acetylcholine have also been shown to be important in AD. However, the mechanism or the exact nature of interactions is still only partially known. Other neurochemical markers such as reduction in N-acetylaspartate and increased myoinositol have been reported (Block et al., 2002; Krishnan et al., 2003; Waldman and Rai, 2003). However, these changes vary and there does not seem to be a consensus on the spatial distribution and therefore studies based on them are still in their preliminary stages.

10.3 Imaging Studies

10.3.1 Anatomical Imaging versus Functional Imaging

The primary reason for the popularity of imaging techniques for detection and diagnosis of AD is the relative ease with which the neurological markers discussed in the previous section can be converted to visual markers. Anatomical changes such as atrophy in the brain or ventricular enlargement can be quantified by visual markers such as changes in volume or shape which are detected using anatomical imaging techniques such as computer tomography (CT) and magnetic resonance imaging (MRI). On the other hand, reduced cerebral activity in affected regions due to neurodegeneration leads to altered cerebral blood flow and biochemistry that are detected using functional imaging techniques such as functional MRI (fMRI), positron emission tomography (PET), single photon emission computed tomography (SPECT), and magnetic resonance spectroscopy (MRS). Recently, another potential marker, diffusion of water, has been identified using diffusion-weighted MRI (Hanyu et al., 1997; Sandson et al., 1999). Microscopic markers such as neurofibrillary tangles and senile plaque cannot be visualized directly using imaging techniques. However, the presence of these structures is associated with secondary effects such as neuronal death, tissue atrophy, and reduced perfusion which can be measured (Xanthakos et al., 1996; Petrella et al., 2003). As an alternative, some attempts have been made to model the aggregation-disaggregation dynamics of plaque formation (Cruz et al., 1997; Urbanc et al., 1999a,b).

Imaging and analytical techniques are used only as aids for diagnosis of AD. A neurologist cannot confirm the presence or absence of the disease from a brain scan alone because even the current state-of-the-art techniques lack the required specificity to AD. One reason for this is the overlap between the symptomatology of AD with other diseases, in which case, improving the technology would not increase the accuracy of diagnosis of AD. Another reason for low specificity is probable shortcomings in the imaging techniques, and improving the image quality can potentially result in greater accuracy and increased ability to distinguish between symptoms of AD and other neuropathologies.

Compared to functional imaging methods, the advantages of anatomical imaging include relative ease of use and low cost. However, neurological markers detected by anatomical imaging modalities are of the gross anatomical type and become noticeable only in advanced stages of AD due to limitations of resolution and existence of artifacts. On the other hand, functional imaging modalities demonstrate greater sensitivity, especially for detecting subtle changes in the earlier stages. However, so far, no single modality has emerged as a preferred method based on diagnostic accuracy, at least in the early stages of AD. All modalities suffer from a common drawback - noise.

Various noise reduction methods based on novel techniques such as wavelet transforms (Adeli et al., 2003; Turkheimer et al., 2003) and artificial neural networks or ANNs (Adeli and Hung, 1995; El Fakhri et al., 2001) have been proposed to resolve this problem.

10.3.2   Identification of Region of Interest (ROI)

Although these imaging techniques are based on different modalities, the basic approach to image analysis is the same and most studies rely on visual image inspection. Different image parameters of the region of interest (ROI) in an image such as area, shape, pixel intensity, and color represent different physiological characteristics of the ROI. For instance, area in the image is a measure of physical size, whereas intensity could be a measure of dynamic properties such as flow, volume, and biochemical concentration. Also, area representation using different imaging modalities has different implications. For instance, in an MRI scan, area represents the physical size of a particular type of tissue whereas in a SPECT or fMRI scan, area represents the area of increased perfusion due to cerebral blood flow. Visual inspection may be performed by a trained radiologist/neurologist quickly but often is biased depending on the examiner. To reduce this bias, attempts have been made to quantify selected characteristics of the ROI in the images for automated or semi-automated image analysis and diagnosis of AD. These attempts are reviewed in the following paragraphs.

For the analysis of any image, first the ROI has to be identified. This requires segmentation algorithms which divide the image area into segments based on some common characteristic of proximity such as location, color, texture, pixel intensity, and brightness or any combination of these. Subsequently, the characteristics of a segment can be quantified and used as a marker of the disease. Segmentation can also be achieved by detecting the contrast-based boundaries between segments using edge detection techniques. A very common method uses threshold-based boundary detection in areas where there is a high contrast. Areas of low contrast are demarcated usually by visual inspection (Jack et al., 1990). Novel methods have been researched for image segmentation as applied to medical images. Good results have been reported with clustering techniques such as standard statistical cluster analysis (Burton et al., 2002), wavelet-based clustering (Barra and Boire, 2000), and adaptive fuzzy c-means clustering (Liew and Yan, 2003). Statistical pattern recognition methods (Andersen et al., 2002) have also been devised for segmentation purposes. Improved segmentation based on pattern recognition methods using supervised and unsupervised ANNs was observed in MRIs (Reddick et al., 1997; Deng et al., 1999; Perez de Alejo et al., 2003) and SPECT scans (Hamilton et al., 1997). Grau et al. (2004) proposed an improved watershed transform, purportedly more accurate for image segmentation. Once the ROI has been segmented, the relevant characteristics of the ROI can be quantified depending on the parameter being measured.

10.3.3   Image Registration Techniques

A major hindrance to accurate comparison of brain scans or parameter quantification is the high potential for variability due to various factors. First, brain scans show significant inter-subject differences because of the physical variations between the brains of any two individuals. Second, even for the same individual, two brain scans can be different because of spatial factors such as position, rotation, and angle of inclination of the head. Finally, inherent flaws in the imaging modalities introduce geometric distortions in the images which can also affect accuracy of the analysis. Therefore, comparison of brain scans to identify changes requires that different brains be mapped onto a standard or control template using scaling and spatial alignment (normalization or registration) techniques (Ashburner et al., 1998; Slomka et al., 2001; Hill et al., 2001).

Image registration relates the spatial location of features in one image to that of the corresponding features in another image (Hill et al., 2001). A standard coordinate system is achieved by using either external markers such as skull screws, stereotactic frames, dental adapters, and skin markers or internal information present in the image. Additional analysis may be performed with intensity values at corresponding locations in the images. Image registration may be achieved by using different types of transformation functions such as rigid, affine, projective, and curved function (Viergever et al., 2001). A similarity function (a function of the ROI volume of the subject, standard ROI volume, and the transformation function) such as robust least squares, mutual information, or count difference is used to assess the success of the registration. The transformed images form the basis for an accurate 2D or 3D comparison of markers such as atrophy or altered cerebral blood perfusion (Radau et al., 2001).

10.3.4 Linear and Area Measures

Early researchers used changes in linear and area measures based on the 2D cross-sectional scans to quantify atrophy. The goal is to find an effective measure that would eliminate the need for more expensive volumetric and functional studies. Attempts have been made to quantify atrophy of different regions of the brain for automated detection and diagnosis of AD (Frisoni et al., 1996). Linear measures of regional frontal atrophy include the *bifrontal index* (defined as the ratio of the maximum distance between the tips of the anterior horns of the lateral ventricles to the width of the brain at the same level) and maximum interhemispheric fissure width. Medial temporal lobe atrophy has been quantified using linear measures such as the *interuncal distance*, defined as the distance between the unci of the temporal lobes (Dahlbeck et al., 1991; Doraiswamy et al., 1993; Early et al., 1993), and minimum thickness and width of the medial temporal lobe (Gao et al., 2003).

Hippocampal atrophy has been measured using the maximum height and width of the hippocampal formation (Erkinjuntti et al., 1993; Huesgen et al., 1993), mean hippocampal angle (Gao et al., 2003), width of the choroid fissure, and width of the temporal horn (Frisoni et al., 2002). Area measures of the temporal horn, interhemispheric fissure, and lateral ventricles (Desphande et al., 2004) have also been employed as measures of atrophy. Different combinations of these linear and area measures such as ratios of the dimensions of anterior and posterior parts of the same feature have also been studied as attempts to quantify changes in shape. Similar measures have been used to quantify dynamic/functional markers such as area of altered perfusion or biochemistry (Mattia et al., 2003).

10.3.5   Volumetric Measures

Most imaging techniques yield 2D scans of the cross-section (also known as slices) of the brain at specified locations. Since linear and area based parameters do not always correlate to an accurate estimate of the volume of the structure, volumetric studies are performed, which is more time consuming. A spatial sequence of slices along a given direction is required to reconstruct a 3D image of the entire brain or a selected part of the brain. A common method for measuring volumes of irregular structures is the point counting method. The points are dubbed *voxel*, similar to pixel for an area. Since the slices are at discrete intervals it is usually not possible to directly count voxels. Instead, the volume is approximated from the number of ROI pixels in all relevant slices. The number of pixels in the ROI is multiplied by the area of a pixel to obtain an approximation of the area of the ROI. The area is then multiplied by the slice thickness to obtain the volume of the ROI in one slice (Bottino et al., 2002; Hampel et al., 2002). Subsequently, the volumes of the ROI in all relevant slices are summed to get the total volume of the ROI. Whole brain volume, cerebrospinal fluid volume, volumes of the temporal, frontal, and parietal lobes, the cerebellum, corpus callosum, and the amygdala-hippocampus complex and their ratios to measure relative changes have been used by different researchers as markers of AD.

However, the boundaries of the aforementioned ROIs may not be clearly defined. In such cases, voxel-based morphometry is used to quantify differences in two images based on a voxel by voxel comparison (Ashburner and Friston, 2000; Rombouts et al., 2000; Baron et al., 2001; Burton et al., 2002; Karas et al., 2003). Other methods such as deformation-based morphometry based on changes in relative positions of features (Ashburner et al., 1998; Janke et al., 2001; Thompson et al., 2004) and tensor-based morphometry that identifies local differences in the shape of brain structures (Ashburner and Friston, 2000) have also been employed with limited success. These methods applied to sequential images can potentially provide insight into the changes associated with the progression of AD. Another method, voxel-compression mapping (Fox et al., 2001), has been developed to view neuronal degeneration or shrinkage over time. However, all these methods are still in their early experimental stages and their clinical value has yet to be established.

10.4   Classification Models

There is still no consensus about the specificity or sensitivity of markers based on imaging or EEG studies. Due to the complex nature of AD and overlap of the symptomatology with other neurological disorders, a single marker may not be sufficient for automated detection or diagnosis of the disease. Additionally, some markers are inherently complex and a simple threshold cannot characterize the disease accurately. Diagnosis of the disease from imaging or EEG studies involves detection of abnormalities by comparison with a healthy human brain. In addition to these inter-individual comparisons, comparisons can be longitudinal in which markers of AD are examined at various stages of life of a patient with potential AD to quantify the rate of progression of the disease. The immense potential for variability not only between individuals but also within the same individual coupled with the lack of defined boundaries between normal and abnormal make this classification problem very difficult. This difficulty becomes even more pronounced when the objective is to design accurate automated classification algorithms for early detection of the disease. Early detection involves capturing very minute changes which may lead to AD. However, so far, even visual inspection by a highly trained neurologist cannot achieve this conclusively.

Due to the wide variety of markers and parameters, this problem may be approached as a pattern recognition problem where the pattern to be recognized may be a spatial, temporal, or spatio-temporal sequence of neuronal firing. Alternatively, the pattern could be that of cortical atrophy as observed in imaging studies. Traditional methods of statistical analysis of significance have long been applied to prove that significant differences exist between patients with probable AD and normal healthy control subjects. Most imaging and EEG studies on the subject of markers of AD employ parametric statistical tests (such as t-test and analysis of variance) to identify the difference. Recent imaging studies have used k-means cluster and k nearest neighbor analysis (Benvenuto et al., 2002) for distinguishing between the different groups of test subjects. Studies have reported slightly better results with images preprocessed with principal least squares analysis than with principal component analysis (PCA) when using a classifier based on linear discriminant analysis (LDA) (Higdon et al., 2004). Besthorn et al. (1997) report that LDA of EEGs yields more accurate classification compared with k-means cluster analysis.

A few applications of ANNs as a classification algorithm have also been reported during the past decade. Pizzi et al. (1995) compared the accuracy of three types of ANNs - backpropagation network, fuzzy backpropagation network, and radial basis function neural network - and concluded that all three types demonstrated greater accuracy than LDA when applied to infrared spectroscopic images preprocessed with PCA. However, only the radial basis function neural network showed the same accuracy when applied to images not preprocessed with PCA. Good classification results were also reported by deFigueiredo et al. (1995) using an optimal interpolative neural network on SPECT images and Warkentin et al. (2004) using a backpropagation network on cerebral blood flow values obtained from patients by the ^{133}Xe (xenon-133 radio isotope) inhalation procedure.

ANNs have also been used to discriminate EEGs of patients with AD from those of normal subjects. Studies have reported that ANNs yield greater classification accuracy than traditional methods such as statistical, clustering, and discriminant analysis (Anderer et al., 1994; Besthorn et al., 1997). ANN classification accuracy is further increased if the EEGs are preprocessed with chaos analysis (Pritchard et al., 1994) or wavelet analysis (Polikar et al., 1997; Petrosian et al., 2000b, 2001). A combination of methods such as chaos analysis, PCA, and discriminant analysis has been reported to yield comparable or under certain conditions better results compared with those obtained using ANNs (Besthorn et al., 1997). It is concluded that using a mixture of markers and a combination of computational techniques can increase the accuracy of algorithms for automated detection and diagnosis of AD. EEG studies are discussed in detail in Chapter 11.

10.5   Neural Models of Memory and Alzheimer's Disease

In addition to the application of ANNs to imaging and EEG studies for classification and decision making, neural modeling has been used to study the dynamics of the human brain and its dysfunction in various neurological disorders such as AD. In order to understand the effects of AD on memory function, neural models are developed to simulate various properties of memory. These neural models are subjected to various trigger conditions which can potentially cause a breakdown in the model function. If the model dysfunction matches the symptomatology of AD, then it can, in an ideal situation, potentially explain three things: (a) the region of the brain responsible for AD, (b) the trigger conditions responsible for AD, and (c) the mechanism responsible for the spread of AD.

This section describes the current hypothesis about the hippocampal models of memory and their dysfunction in AD. Models of memory dysfunction in AD are henceforth referred to as neural models of progression of AD. These models are based on significantly different approaches to neural modeling as described next.

10.5.1   Approaches to Neural Modeling

Neural models can be categorized as connectionist or biophysical. Both approaches have been used to simulate small scale networks in the brain with the goal of understanding the underlying mechanisms of AD. The *connectionist* neural models employ simplified neural network dynamics that emphasize the importance of synaptic connections and the role of neurons as simple spatio-temporal integrators in associative memory formation (Hasselmo and McClelland, 1999; Finkel, 2000). For this reason, they are used to model large scale processes in the brain. Two specific connectionist neural models of progression of AD have been proposed based on experimental evidence regarding the effects of progression of AD on synaptic connections (Duch, 2000).

The *biophysical* neural models employ detailed biophysical properties of a single neuron (known as a *spiking* neuron). These properties include the physiological effects of neuromodulation such as bursting and spiking. Spiking neurons are designed based on actual experimental data obtained on ion channel properties. The ion channels transduce presynaptic neurotransmitter action into membrane depolarization and repolarization which lead to firing of action potentials (Traub et al., 1991, 1994; Pinsky and Rinzel, 1994; Migliore et al., 1995; Wang and Buzsaki, 1996). These models are interconnected networks of spiking neurons which simulate the aggregation effects of the properties of individual neurons and associated phenomena. For this reason, the biophysical neural models are referred to as *biological analogs* of the connectionist neural models by Menschik et al. (1999). They can be used to simulate memory dysfunction such as that seen in AD by manipulating neuron properties and parameters such as cholinergic neuromodulation, theta and gamma oscillations, diversity of neuron types (for example, interneurons, bursting and spiking pyramidal cells), and input patterns (Menschik and Finkel, 1998, 1999, 2000).

The connectionist neural models do not require as much simulation detail as their biophysical counterparts which makes the latter computationally intensive, especially for large networks. Therefore the biophysical models are only used to model the behavior of a small set of neurons in the brain. Despite the differences, it has been observed experimentally that both the connectionist and biophysical models demonstrate certain common behaviors depending on manipulations of network parameters such as number of neurons, strength of connections, and neuronal response to neurotransmitter actions.

10.5.2   Hippocampal Models of Associative Memory

Various parts of the brain are affected during the late stages of AD. The first changes, however, are noticed in the mesial temporal lobe which includes the hippocampal formation (Finkel, 2000). The hippocampal formation is especially significant for memory function which is impaired in AD. Therefore, studies have focused on hippocampal dysfunction as a possible explanation for AD (Price, 2000). Neural models have been developed to simulate the function of the hippocampus with respect to memory formation. However, no one model is sufficient for this purpose. Rather, different models reproduce different memory properties and therefore it is hypothesized that multiple models implemented simultaneously may result in improved performance (Hasselmo and McClelland, 1999; Duch, 2000; Finkel, 2000).

One model, the attractor neural model, simulates memory properties such as error correction and pattern completion (Hasselmo and McClelland, 1999; Finkel, 2000). This model is based on the assumption that memory corresponds to a stable spatio-temporal pattern or state of activated neurons. In a given network, there may be more than one such stable state corresponding to multiple memories. These stable states are termed fixed point attractors or attractor states. Given a network containing N neurons, any activation state is represented by a point in an N-dimensional feature space. Assuming that the activation state of any neuron can be represented by one of two states (0 or 1), the total number of possible memory states is 2^N. If the network is activated in an initial state similar to the attractor state, then the network gravitates toward the attractor state. Though the final state of the network is always one of the attractor states, the selection of the specific attractor state is dependent on the initial state. Therefore, this model converges to the true memory even when presented with incomplete or partially incorrect information about the memory (Hasselmo and McClelland, 1999; Finkel, 2000).

In the attractor neural model the memory patterns are stored as weights of synaptic connections. Even though this model accurately simulates error correction and pattern recognition, it suffers from a significant drawback with respect to encoding (or storage) of a new input. A new input is also treated as an incomplete or incorrect pattern and therefore gets *contaminated* by previously stored patterns (Finkel, 2000). Hasselmo and McClelland (1999) also comment on the basis of experimental studies that a significantly greater effort is required for memory recall compared with recognition. These observations are not new and have led to the hypothesis that hippocampal processing switches to different modes during the performance of different functions such as encoding and recall (Finkel, 2000). The mode switching neural model is an extension of the attractor neural model which tries to model the robustness of memory with respect to storage of new memories without compromising the advantages of the previous model.

10.5.3   Neural Models of Progression of AD

In AD, cell death is normally limited to no more than about 10% of the neuronal population which does not account for the corresponding level of cognitive deficit. Experimental observations have led to the conclusion that the primary factor responsible is a 50% decrease in the number of synaptic connections often represented by the total synaptic area per unit volume (Finkel, 2000). This phenomenon is termed *synaptic deletion*. The brain attempts to compensate for the loss of synaptic connections by increasing the strength of the remaining synaptic connections. This strategy, termed *synaptic compensation*, is successful in limiting the cognitive deficits in the initial stages of AD. However, in more advanced stages, the loss of synapses is too much to be overcome by neuromodulation and therefore results in an increasing severity of cognitive deficits. Synaptic deletion and compensation form the basis of the *synaptic deletion and compensation model* (Horn et al., 1993, 1996; Ruppin and Reggia, 1995; Reggia et al., 1997). Using this model it has been demonstrated that, in a Hopfield type ANN architecture, the loss of synaptic connections causes loss of memory and distortion of learned patterns. However, the rate of memory deterioration is significantly reduced by increasing the strength of the remaining synaptic connections (or weights) by a constant multiplicative factor.

Another phenomenon observed in associative networks is *runaway synaptic modification* (Hasselmo, 1994, 1995; Siegle and Hasselmo, 2002). In associative models of memory, storage of one memory as a spatio-temporal neuronal activation pattern is associated with the storage of other related memories. Storage of a new memory activates similar patterns which may interfere with previous associations if (1) there is overlap between patterns or (2) the memory capacity is exceeded (Hasselmo, 1994). This interference results in a significant increase in the number of associations that are stored by the network. This increase often leads to a pathological increase in the strength of synaptic connections which results in increased neuronal activity, high metabolic demands, and eventually cell death. This phenomenon is referred to as excitotoxicity.

The *runaway synaptic modification model* has also been used to demonstrate that separate mechanisms exist in the brain for memory encoding and recall in order to minimize the interference-induced excitotoxicity. According to this hypothesis, inhibitory and excitatory cholinergic neuromodulation is the primary factor that controls the switching from one mode to the other. In normal states, neuromodulation is sufficient to prevent runaway synaptic modification. However, in pathological states (such as AD where cholinergic neuromodulation is impaired) or beyond a certain level of memory overload in healthy patients, runaway synaptic modification is inevitable (Hasselmo,

1994; Reggia et al., 1997; Duch, 2000). The threshold level for AD patients, however, is substantially lower than that for healthy subjects.

These neural models of memory and progression of AD have been used in various combinations to explain the mechanisms underlying AD. Even though current knowledge on the topic is constantly evolving, no mechanism has been conclusively proposed as a complete explanation of the subject.

11

Alzheimer's Disease: Models of Computation and Analysis of EEGs

---

11.1 EEGs for Diagnosis and Detection of Alzheimer's Disease

The development of anatomical and functional imaging modalities such as computer tomography (CT), magnetic resonance imaging (MRI), functional MRI (fMRI), and positron emission tomography (PET) has contributed significantly to Alzheimer's disease (AD) research and the understanding of the disorder. Although nothing conclusive has been established, imaging techniques have gained popularity due to the relative ease with which neurological markers such as neuronal loss, atrophy of brain tissue, and reduced blood perfusion can be converted to visual markers on brain images (Xanthakos et al., 1996; Petrella et al., 2003).

Due to the expense of specialized experts and equipment involved in the use of imaging techniques, a subject of significant research interest is detecting markers in EEGs obtained from AD patients. EEG studies are non-invasive and, in the case of AD, show comparable sensitivity and improved specificity compared to imaging studies (Bennys et al., 2001; Benvenuto et al., 2002). Since AD is a dysfunction of the cerebral cortex, abnormalities in field potentials (recorded by EEGs) in the cortex can be directly correlated to the pathological changes in the structure and function of the cortical layers in AD (Jeong, 2004). Additionally, longitudinal studies (along the lives of patients with probable AD) have shown that the degree of abnormality associated with AD has been found to be directly proportional to the degree of progression of the disease (Coben et al., 1985; Dierks et al., 1991; Soininen et al., 1991; Prichep et al., 1994).

The most commonly used markers of abnormality in EEGs from AD patients are based on time-frequency measures such as frequency and correlation (representing the statistical similarity between two EEGs over time) which have been extended to include spatial changes between different regions of the brain. Recent research in this area is based on novel analytical techniques that can be classified into the following categories: (a) time-frequency analysis, (b) wavelet analysis, and (c) chaos analysis.

11.2   Time-Frequency Analysis

The most common marker of abnormality in EEGs seen in AD is a decrease in the high frequency content of the EEG termed *EEG slowing*. EEG slowing results from an increase in the power in low frequency bands and a decrease in the power in the high frequency bands of the EEG. Power of a frequency band is representative of the activity in that frequency band and is calculated using the Fourier transform of the signal. Studies have established changes in the four primary frequency bands of the EEG: *alpha* (8-12 Hz) (normal), *beta* (13-30 Hz) (generally seen in anxiety states or as a result of medication effects), *delta* (0-4 Hz), and *theta* (4-7 Hz) (pathological rhythms) at various stages of the disease (Coben et al., 1985; Soininen et al., 1992; Miyauchi et al.,

1994; Besthorn et al., 1997; Wada et al., 1997; Pucci et al., 1999; Huang et al., 2000; Bennys et al., 2001; Stevens et al., 2001; Jeong, 2004). Earlier stages of AD are characterized by an increase in *theta* activity and a decrease in *beta* activity. As the disease progresses, these changes are followed by a decrease in *alpha* activity. Increase in *delta* activity occurs at more advanced stages of the disease. In this context, increase in activity is defined as an increase in the absolute power in the frequency band in the EEG.

Recent studies have found that EEG slowing is more prominent in rapid eye movement (REM) sleep than in wakefulness (Montplaisir et al., 1998; Musha et al., 2002). The EEG slowing of different parts of the brain also display significant differences (Pucci et al., 1999). Studies report increased EEG slowing in the right postero-temporal region (Duffy et al., 1984), the occipital region (Soininen et al., 1992), and the left frontal and temporal lobes (Miyauchi et al., 1994). Wada et al. (1997) have found increased *delta* activity in the frontal lobe and increased *theta* activity in the right parietal and postero-temporal regions. They also confirm that, unlike normal controls, AD patients lack a predominance of *alpha* activity in the posterior regions.

The spectral analysis of the various frequency bands is commonly performed using fast Fourier transform (FFT) on artifact-free EEGs. In order to study the EEGs in greater detail, sometimes the frequency band *beta* is examined as two separate sub-bands, $beta_1$ (13-21 Hz) and $beta_2$ (21-30 Hz). The same procedure is followed for the other bands if required. Different linear combinations and ratios of the absolute power values for the four frequency bands and their sub-bands have been employed as measures to detect EEG slowing. Gueguen et al. (1991) employ the mean dominant frequency and the ratio *alpha/theta*. Bennys et al. (2001) report increased accuracy with the combinations $(theta/alpha) + beta_1$ and $delta + (theta/alpha) + beta$. Alternatively, other measures such as relative power, defined as the ratio of absolute power for a frequency band to the total power for all four bands (Coben et al., 1985; Leuchter et al., 1993; Stevens et al., 2001), and EEG spectral profiles derived from spectral density plots (Signorino et al., 1995) have also been suggested as indicators of EEG slowing.

Attempts have been made recently to come up with improved measures that can detect and diagnose AD. Huang et al. (2000) argue that interpretation of FFT analysis is affected by the choice of reference electrode and variability in a large number of physiological parameters. In order to reduce this variability, they suggest using the FFT-dipole approximation method (Lehmann and Michel, 1990; Michel et al., 1993) that represents multi-channel EEG data by one oscillating dipole source (as a function of phase angles) in the frequency domain. The dipole source for any frequency band represents the centroid for the brain electrical activity in that band. Under controlled conditions, the conventional FFT is just as accurate as this method but Huang et al. (2000) argue that it may perform better in a random patient population. Musha et al. (2002) report satisfactory results using another measure termed *mean alpha dipolarity* which is defined as the electric current dipole distribution of the *alpha* band in the cortex.

Another marker of abnormality that has been studied but not as widely as EEG slowing is reduced *EEG coherence* which implies reduced corticocortical connectivity (Besthorn et al., 1994; Jeong, 2004). EEG coherence is defined as the spectral correlation between two spatially distributed EEG signals for a given frequency band. EEG coherence can be local or global. Local coherence is a measure of the local differences in dynamics between two cortical areas and is computed as the pairwise correlation between EEGs from two electrodes. On the other hand, global coherence is a measure of average brain dynamics and is computed as the average of all the local coherence values (Stevens et al., 2001). However, coherence is sensitive to the distance between the electrodes and therefore, for meaningful comparisons, it should only be computed for immediately adjacent electrodes. This is a disadvantage for research involving study of the changes in corticocortical connectivity between distant parts of the brain in AD.

The selection of the *immediately adjacent* electrode out of all the surrounding electrodes is also important. Besthorn et al. (1994) propose a *spatially averaged coherence* scheme in which the EEG coherence for an electrode is equal to the average of the coherence values obtained from pairwise comparisons of that electrode with all electrodes immediately adjacent to it. Dunkin et al. (1995) perform a layout-based selection in which one electrode is compared to the next one in a specific pattern. With respect to coherence studies, Besthorn et al. (1994) have found significant differences in the *alpha*, *beta*, and *theta* frequency bands, whereas other investigators report significant differences in either the *alpha* (Locatelli et al., 1998) or *theta* band (Stevens et al., 2001). Berendse et al. (2000) suggest using the same techniques with magnetoencephalograms (MEGs) instead of EEGs because MEGs are less labor intensive and yield comparable results. Additionally, it is much faster to obtain MEGs with higher resolution for detecting small local changes due to a large number of built-in detection coils. Comparable resolution for EEGs may be achieved by employing a large number of electrodes which is very time intensive.

Leuchter et al. (1994) suggest a measure dubbed *cordance* that combines the absolute and relative power measures. For instance, low absolute but high relative power yields a negative value and is termed discordance. Similarly, a positive value of cordance is termed concordance. Based on comparisons with imaging studies they suggest that cordance is sensitive to cortical deafferentation (loss of afferent inputs). Cook and Leuchter (1996) suggest that cordance and coherence be used simultaneously to detect synaptic dysfunction in AD.

Additionally, the photic driving response to light stimulation decreases in AD patients possibly due to lesions in the occipital regions (Signorino et al., 1996; Wada et al., 1997). To quantify the response of a frequency band, Signorino et al. (1996) propose a measure termed *power index*, defined as the percentage change in absolute power in the band due to eye opening. Henderson et al. (2002) employ a simple metric based on zero-crossing intervals in EEGs and report good classification results. Bispectral analysis has also been used to measure the changes in phase coupled frequencies (Villa et al., 2000). Simeoni and Mills (2003) propose a bispectral EEG analysis based on global coherence and second-order phase coupling between the four EEG frequency bands to quantify the progression of AD but do not report any conclusive evidence of differences.

11.3   Wavelet Analysis

Most of the research involving time-frequency analysis of EEGs has focused on standard FFT-based methods. Wavelet-based methods have some inherent advantages over FFTs as described in Chapter 2 especially for applications involving non-stationary transient signals such as EEGs. Recently, wavelet-based filtering and feature extraction strategies applied to other neuropathologies such as epilepsy have shown promising results (Adeli et al., 2003, 2005b; Petrosian et al., 1996, 2000a).

Very little work has been reported on wavelet transforms applied to EEG in AD. Petrosian et al. (2000b, 2001) employ the second order Daubechies wavelet to perform a multilevel decomposition of the EEG. The high frequency components at every level along with the original EEG are input into a recurrent neural network classifier for classification as AD or healthy. They report that the wavelet-based preprocessing (high-pass filtering) step improves the classification accuracy as compared to classical methods. A method closely related to the EEG, called evoked potential or event related potential (ERP) has also been used to study AD. ERPs are segments of EEGs that are clearly identified as resulting from the perception of external stimuli (such as light or sound) and the subject's response to it. Polikar et al. (1997) suggest a similar multilevel decomposition of ERPs and a subsequent neural network classifier. However, the neural network is trained with a subset of the wavelet coefficients (selected on the basis of highest amplitudes) to reduce the size of the feature space and increase the robustness of the algorithm. Ademoglu et al. (1997) employ spline wavelets in an attempt to accurately identify features associated with abnormalities in the latency and amplitude of the P100 component of the ERP (referring to a positive peak observed 100 ms after photic stimulation).

11.4 Chaos Analysis

In the last decade, a lot of interest has developed in studying the nonlinear dynamics of the brain by means of chaos analysis of EEGs. Chaos analysis can quantify certain characteristics of the EEG that are not readily visible to a naked-eye examination. Toward this end, studies have been performed on EEGs obtained from both (a) normal states of the brain such as sleep (Molnar and Skinner, 1991; Roschke and Aldenhoff, 1991; Niestroj et al., 1995; Zhang et al., 2001; Kobayashi et al., 2001, 2002; Ferri et al., 1998, 2002, 2003; Shen et al., 2003) and meditation (Aftanas and Golocheikine, 2002; Efremova and Kulikov, 2002), as well as (b) pathological states such as schizophrenia (Roschke and Aldenhoff, 1993; Paulus et al., 1996; Huber et al., 1999, 2000;

Paulus and Braff, 2003) and epilepsy (Iasemidis and Sackellares, 1991; Bull-more et al., 1992; Iasemidis et al., 1994; Lopes da Silva et al., 1994; Elger and Lehnertz, 1994, 1998; Hively et al., 1999; Andrzejak et al., 2001; Litt and Echauz, 2002; Adeli et al., 2007; Ghosh-Dastidar et al., 2007).

A marker of abnormality similar to local coherence, called *mutual information* (MI) has also been studied as a measure of corticocortical connectivity (Jeong et al., 2001b; Jeong, 2002). MI can be of two types: cross mutual information (CMI) and auto mutual information (AMI). Coherence is based on the correlation function which measures linear dependence of EEGs from two different electrodes whereas CMI is based on the MI function which measures linear and nonlinear dependence. AMI is also based on the MI function but it is a measure of the dependencies between a single-channel EEG and the same EEG with some specified time delay. Jeong et al. (2001b) report a reduction of inter-hemispheric CMI over frontal and antero-temporal regions suggesting loss of corticocortical connections in AD. However, this reduction may be a result of sensitivity to distance (as explained in the case of coherence).

Many researchers has focused on quantitative measures of *EEG complexity*, another marker of abnormality. The most commonly used measure of complexity is the dimension of the attractor, in particular, the correlation dimension, CD (Takens, 1981; Williams, 1997; Borovkova et al., 1999; Jiang and Adeli, 2003). It is usually computed using a single-channel EEG and the same EEG with specified time delays, called time-delay embedding (Takens, 1981), to quantify the temporal evolution of the EEG from a specific location in the brain. A high value of the dimension implies higher complexity. Studies have claimed that the value of CD increases in mental states such as wakefulness and cognitive task performance (Stam et al., 1996; Meyer-Lindenberg et al., 1998; Molle et al., 1999; Jelles et al., 1999). CD also increases with increased intelligence (with age) from adolescence to adulthood (Anokhin et al., 1996;

Meyer-Lindenberg, 1996). Reduction in CD is caused by factors that cause cognitive deficits such as sleep deprivation (Jeong et al., 2001c) and epileptic seizures (Elger and Lehnertz, 1994, 1998; Adeli et al., 2007; Ghosh-Dastidar et al., 2007). AD is also associated with neuronal loss which leads to cognitive deficits and therefore it is not surprising that EEGs in AD patients show lower complexity or reduced CD values (Pritchard et al., 1994; Besthorn et al., 1995; Jelles et al., 1999; Jeong et al., 1998, 2001a). CD values obtained from MEGs are significantly reduced in the *delta* and *theta* bands in AD (van Cappellen van Walsum et al., 2003). Additionally, the responsiveness of CD to photic stimulation is reported to be reduced mildly or non-existent in AD patients (Pritchard et al., 1994).

Studies have also been performed using other measures of complexity. Pritchard et al. (1994) examine a parameter, *saturation correlation*, defined as the saturation value of the Pearson correlation computed incrementally for estimated embedding dimensions (Bullmore et al., 1992; Cao, 1997; Jiang and Adeli, 2003; Notley and Elliott, 2003; Natarajan et al., 2004). They found that in AD patients the saturation correlation is low in the frontal midline regions of the brain whereas in normal controls it is high in the frontal midline regions. Jeong et al. (1998) report reduced values of the largest Lyapunov exponent (Wolf et al., 1985; Rosenstein et al., 1993; Williams, 1997; Hilborn, 2001) in AD patients. Jeong et al. (2001b) employ the rate of saturation of the AMI with incremental time delays as a measure of complexity and report that this rate is lower in patients with AD, implying a lower complexity as compared to normal controls. Woyshville and Calabrese (1994) have found a significant reduction in the fractal dimension in patients with AD. This hypothesis has been questioned by Henderson et al. (2002) based on the possibility that normal human EEG is non-fractal in nature.

In addition to the usual definition of complexity based on the temporal evolution of the EEG, the concept of *global complexity* has been introduced to study the spatio-temporal evolution of the EEG (Dvorak, 1990; Wackermann et al., 1993). Instead of time-delay embedding applied to a single-channel EEG, global complexity is based on spatial embedding on multi-channel EEGs. The measures of global complexity are computed similarly to the measures of complexity mentioned above. However, the new measures have the word *global* added. For instance, correlation dimension is a measure of complexity and global correlation dimension is a measure of global complexity. The changes in global complexity have been studied in various mental states such as drowsiness and sleep (Matousek et al., 1995; Szelenberger et al., 1996a,b; Sulimov and Maragei, 2003), increased cognition under the influence of a nootropic drug (a drug that enhances cognition, memory, and learning) (Wackermann et al., 1993; Kondakor et al., 1999), schizophrenia (Saito et al., 1998; Lee et al., 2001), and hypnosis (Isotani et al., 2001).

In patients with AD, studies report reduced values for the global correlation dimension and the global largest Lyapunov exponent but no significant differences were found in the values of the global Kolmogorov-Sinai entropy (Stam et al., 1994, 1995; Yagyu et al., 1997). Kim et al. (2001) propose the use of eigenvalue distribution using Karhunen-Loeve decomposition to distinguish between AD patients and normal controls. They report that in severe AD the three largest eigenvalues are much larger than the rest and consequently the eigenvalue distribution decays rapidly whereas in normal controls the decay is smoother. Despite some good results, there does not seem to be a consensus as to which of the two markers of abnormality, coherence or global coherence, models the dynamics of the EEG more accurately (Pritchard et al., 1996; Pezard et al., 1999; Pritchard, 1999; Wackermann, 1999).

Another measure, termed *neural complexity* has been proposed to quantify the relationship between functional segregation and integration in the brain

(Tononi et al., 1994). This measure is an attempt toward explaining how the brain binds together information processed by different functional modalities. They hypothesize that the neural complexity is higher during the performance of cognitive tasks (Tononi et al., 1996; Tononi and Edelman, 1998; Sporns et al., 2000) such as perception of 3D images from random dot stereograms (Tononi et al., 1998). This hypothesis is further supported by more recent research (Burgess et al., 2003; Gu et al., 2003). Based on this argument, the model predicts that in neuropathologies associated with cognitive deficits, the neural complexity is reduced due to a decreased capability to process information. While this happens to be the consensus among researchers so far, the research by van Putten and Stam (2001) on human EEGs indicates the opposite. They report an increase in neural complexity values during epileptic seizure and hypoxic trauma (brain damage due to reduced oxygen supply) when cognition is severely impaired. van Cappellen van Walsum et al. (2003) report neural complexity values from broad frequency spectrum MEGs taken from AD patients yield no significant differences compared to normal controls but when the MEGs are split up into their four primary frequency bands, the *alpha*, *delta*, and *theta* bands show increased neural complexity in the case of AD patients.

A major concern in application of chaos theory to EEGs is the validity of the hypothesis that the measures of complexity are quantifiers of nonlinear dynamics (Jelles et al., 1999; Jeong, 2002, 2004). Usually, criteria such as finite correlation dimension and a positive Lyapunov exponent (Williams, 1997) are used to determine if a system contains deterministic chaos or not. However, earlier studies demonstrated that these criteria could also be satisfied by data containing colored noise (defined as noise containing more of specific frequency ranges than others) instead of deterministic chaos (Osborne and Provenzale, 1989; Pijn et al., 1991; Rapp et al., 1993). Attempts have been made to address this concern using surrogate datasets that are created from the original EEGs by the phase randomization method of Theiler et al. (1992). This method preserves the EEG power spectrum but destroys the non-linear structure of the information to ensure that the surrogate dataset does not contain deterministic chaos. Subsequently, measures of complexity are obtained from both the original EEG as well as the surrogate dataset. The values of the measures in the two cases will be significantly different only if the original EEG contains deterministic chaos. However, there is no definite consensus about the presence of deterministic chaos in EEGs as some studies have found no evidence (Theiler et al., 1992; Pritchard et al., 1995; Palus, 1996; Theiler and Rapp, 1996; Pereda et al., 1998; Jeong et al., 1999) while many others have found significant evidence (Rombouts et al., 1995; Stam et al., 1995; Fell et al., 1996; Ehlers et al., 1998; Jelles et al., 1999; Lee et al., 2001). Despite these controversies, the current popularity of chaos analysis stems from its usefulness in the classification of various mental states (as discussed earlier).

11.5   Concluding Remarks

Most of the strategies and techniques for detection and diagnosis of AD discussed in Chapters 10 and 11 are still in their early experimental stages. Researchers have not yet found conclusive evidence regarding the specificity and sensitivity of the neurological markers and diagnostic techniques based on them. Similarly, there seems to be no consensus regarding the various hypotheses of progression of AD from the point of view of different disease states (such as mild cognitive impairment, presenile dementia of the Alzheimer type, and senile dementia of the Alzheimer type) and clear cut boundaries between them. EEG and imaging studies are heavily dependent on the accuracy of the markers selected to quantify the progression of AD. With respect to neural models, simulations have been performed with small numbers of neurons due to computational limitations and accuracy of the results has not been conclusively established.

The problem is further magnified when the design and development of algorithms for automated detection and diagnosis of AD are considered. Automated algorithms require a quantitative parametric representation of the qualitative or visual aspect of markers of AD. Attempts have been made to establish threshold values for these parameters to distinguish disease states from normal states. However, in most cases simple thresholds are insufficient for this purpose. We believe that a combination of computational paradigms such as wavelet transforms, chaos theory, and artificial neural networks should be used to solve the complicated automation problem of detection and diagnosis of AD.

The vast number of physiological parameters involved in the poorly understood processes responsible for AD yields a large combination of parameters that can be manipulated and studied. Many parameters and their combinations have yet to be studied. As was observed in the case of epilepsy discussed in Part II, dynamics that do not show up in full spectrum EEGs may show up in specific sub-bands or some combination of sub-bands. Multiple parameters such as trajectory divergence, entropy, and Lyapunov exponents quantifying the system attractor can be included in the feature space in order to increase the accuracy of classification algorithms.

A single modality of investigation (such as imaging study or EEG study) also may not be sufficient. Instead, a combination of parameters from different investigation modalities seems to be more effective in increasing the accuracy of detection and diagnosis. Finally, in addition to exploring various combinations of currently available modeling techniques, new mathematical models may also be required in order to simulate the dynamics of the brain and provide explanations for cortical processes that cannot be modeled by current mathematical models. A possible application of the wavelet-chaos methodology, presented earlier in Part II, is presented in Chapter 12 for extracting relevant EEG-based features or markers in Alzheimer's disease.

12

A Spatio-Temporal Wavelet-Chaos Methodology for EEG-Based Diagnosis of Alzheimer's Disease

---

12.1 Introduction

As described in the previous chapter, a commonly used marker of abnormality in EEG studies of AD is *EEG slowing*, which quantifies the impairment of the *temporal* or the frequency aspect of the information processing ability of the brain. Reduced values of similarity measures such as *coherence* (or correlation) (Besthorn et al., 1994; Locatelli et al., 1998; Stevens et al., 2001; Jeong, 2004) and *mutual information* (Jeong et al., 2001b) between EEGs recorded from various regions of the brain have also been proposed as potential markers of abnormality. These quantify the impairment of the *corticocortical* connectivity or the *spatial* aspect of the information processing ability of the brain which involves the proper assimilation of information by various brain regions. Together, the spatio-temporal impairment of information processing in the brain can be correlated to the symptoms of memory loss and cognitive impairment in AD.

In the past decade, a different approach has been advocated for extracting potentially more accurate markers of abnormality from EEGs. It has been hypothesized that the complexity of the human brain (directly proportionate to the information processing capability) can be represented by the *complexity* of the non-linear chaotic dynamics underlying the EEG (Pritchard et al., 1991; Ikawa et al., 2000; Abasolo et al., 2005; Escudero et al., 2006). Studies based on a commonly used measure of complexity in chaotic systems, the correlation dimension (CD), have claimed that the value of CD is increased as a result of wakefulness and cognitive task performance (Stam et al., 1996; Meyer-Lindenberg et al., 1998; Molle et al., 1999) and is decreased as a result of factors that cause cognitive deficits such as sleep deprivation and epileptic seizures (Elger and Lehnertz, 1998; Hively et al., 1999; Adeli et al., 2007). A similar reduction in complexity is also observed in AD, where the neuronal loss and reduction in corticocortical connectivity lead to simpler brain dynamics compared to a healthy brain (Pritchard et al., 1994; Besthorn et al., 1995; Jelles et al., 1999). A similar measure, *neural complexity*, has also been proposed to quantify how well the brain binds together information (Tononi et al., 1994, 1998; Sporns et al., 2000). In neuropathologies associated with cognitive deficits, the neural complexity appears to be reduced due to a decreased capability to process information (Burgess et al., 2003; Gu et al., 2003). The largest Lyapunov exponent (LLE), a measure of EEG *chaoticity*, also appears to be reduced, implying lower brain complexity in AD patients (Jeong, 2004). It should be noted that there is a distinction between *brain* complexity and *mathematical* complexity.

Although a decrease in EEG complexity in specific brain regions has been found in AD, no investigations have been reported regarding the localization of the decrease to specific EEG sub-bands. Similar to the findings in the case of epilepsy, the authors hypothesize that the EEG sub-bands may yield more accurate information about underlying neuronal dynamics (Adeli et al., 2007). Changes that are not evident in the original full-spectrum EEG may be amplified when each sub-band is analyzed separately. In fact, the phenomenon of EEG slowing involves changes in the power of specific EEG sub-bands which appears to support this hypothesis.

In this chapter, a wavelet-chaos methodology is presented for spatio-temporal analysis of EEGs and EEG sub-bands for diagnosis of AD. The methodology is applied to two different groups of multi-channel EEGs: (a) healthy subjects and (b) patients with a diagnosis of probable AD collected under two conditions: (a) eyes closed and (b) eyes open. Each EEG is decomposed into the four EEG sub-bands implicated in AD: *delta* (0-4 Hz), *theta* (4-7 Hz), *alpha* (8-12 Hz), and *beta* (13-30 Hz) using wavelet-based filters. The non-linear chaotic dynamics of the original EEGs are quantified in the form of CD and LLE. Similar to the original EEG, each sub-band is also subjected to chaos analysis to investigate the localization of the changes in CD and LLE to specific sub-bands of the EEG. Subsequently, the effectiveness of CD and LLE in differentiating between the two groups is investigated based on statistical significance of the differences. The eyes open and eyes closed conditions are analyzed separately to evaluate the effect of additional information processing in the brain resulting from visual input and attention. EEGs from multiple electrode channels corresponding to multiple loci in the brain are investigated using the wavelet-chaos methodology to discover areas of the brain responsible for or affected by changes in CD and LLE.

12.2   Methodology

12.2.1   Description of the EEG Data

The dataset used to investigate the methodology consists of multi-channel EEGs from two different groups of subjects: healthy elderly subjects (control group) and patients with a diagnosis of probable AD. The control group consists of 7 subjects (average age of 71) with no history of neurological or psychiatric disorder. The AD group consists of 20 subjects (average age of 74) diagnosed with probable AD as per National Institute of Neurological and Communicative Disorders and Stroke-Alzheimer's Disease and Related Disorder Association (NINCDS-ADRDA) and Diagnostic and Statistical Manual of Mental Disorders, Third Edition, Revised (DSM-III-R) criteria. The EEGs are recorded using 19 electrodes in the standard 10-20 configuration, shown in Fig. 12.1, with forehead as ground and linked mandibles as reference at a sampling rate of 128 Hz. Each channel of data is recorded as the potential difference between an electrode and a reference electrode.

For both groups, the EEGs are collected under two conditions: eyes open and eyes closed. Eight-second EEG segments free from eye blink, motion, and myogenic artifacts are extracted from the EEG recordings. The numbers of such 8-second EEGs for each subject in the control and AD groups are 1 (total of $7 \times 1 = 7$) and 4 (total of $20 \times 4 = 80$), respectively. The EEGs obtained had previously been band-limited to the range of 1-30 Hz during the EEG recording (online) and preprocessing (offline) stages and will henceforth be referred to simply as the EEG. The range is sufficient to extract the four EEG sub-bands implicated in AD: *delta*, *theta*, *alpha*, and *beta*. The reader should refer to Pritchard et al. (1991) for further data acquisition details on a similar but smaller dataset.

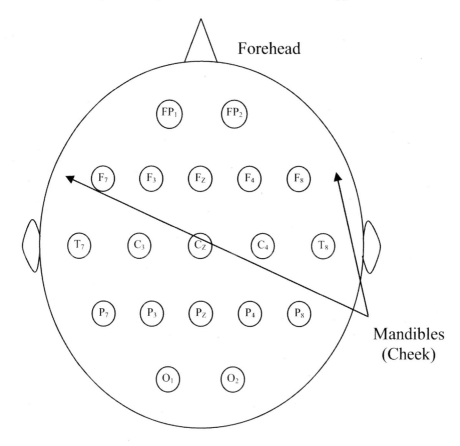

FIGURE 12.1

Electrode loci in the modified 10-20 electrode configuration for 19-channel EEG

The electrode configuration shown in Fig. 12.1 is the same as that shown in Fig. 5.2 except for the renaming of four electrodes as per the recommendations of the American Clinical Neurophysiology Society. To better conform to the naming conventions (with respect to the corresponding lobes of the brain), the T_3, T_5, T_4, and T_6 electrodes were renamed T_7, P_7, T_8, and P_8, respectively. This modified configuration is used in this chapter.

12.2.2   Wavelet Decomposition of EEG into Sub-Bands

To obtain the four EEG sub-bands, the EEG signal is decomposed into progressively finer details by means of multi-resolution wavelet analysis (as described in Chapter 2). The EEG is subjected to a level 3 decomposition using fourth order Daubechies wavelet transform, as shown in Fig. 12.2. After the first level of decomposition, the band-limited EEG (1-30 Hz), denoted by **s** in Fig. 12.2, is decomposed into its higher resolution, $\mathbf{d}_1$ (15-30 Hz), and lower resolution, $\mathbf{a}_1$ (1-15 Hz), components. In the second level of decomposition, the $\mathbf{a}_1$ components are further decomposed into higher resolution, $\mathbf{d}_2$ (8-15 Hz), and lower resolution, $\mathbf{a}_2$ (1-8 Hz), components. Following this process, after three levels of decomposition, the components retained are $\mathbf{a}_3$ (1-4 Hz), $\mathbf{d}_3$ (4-8 Hz), $\mathbf{d}_2$ (8-15 Hz), and $\mathbf{d}_1$ (15-30 Hz). Reconstructions of these four components using the inverse wavelet transform approximately correspond to the four physiological EEG sub-bands *delta*, *theta*, *alpha*, and *beta*. Minor differences in the boundaries between these components and the boundaries between the EEG sub-bands are of little consequence due to the physiologically approximate nature of the sub-bands.

12.2.3   Chaos Analysis and ANOVA Design

The chaos methodology is procedurally similar to that described in Chapter 7. The 19-channel EEGs collected under 2 conditions (eyes open and eyes closed) from 7 healthy subjects and 20 AD patients yield a total of $19 \times 2 \times (7 + 20) = 1026$ EEGs. Similar to the original EEG, each sub-band is also subjected to chaos analysis to investigate the localization of the changes in CD and LLE to specific sub-bands of the EEG. As a result of the wavelet-chaos methodology (Fig. 12.2), each EEG is quantified by ten parameters: CD, LLE, δCD, δLLE, θCD, θLLE, αCD, αLLE, βCD, and βLLE. In this notation, first described in Chapter 8, the parameter prefix denotes the EEG

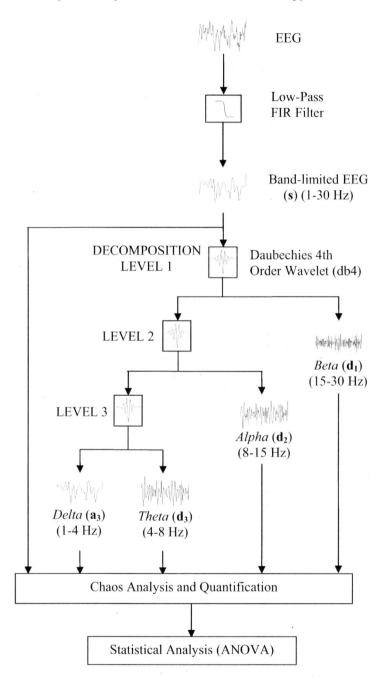

FIGURE 12.2
Overview of the EEG preprocessing and wavelet decomposition for sub-band
extraction prior to chaos quantification and statistical analysis

sub-band from which the parameter is computed. Absence of a prefix indicates that the parameter is computed from the band-limited EEG. The five CD-based parameters represent the complexity and the five LLE-based parameters represent the chaoticity of the EEG and EEG sub-bands.

The statistical investigation is performed in three steps. First, for each of the ten parameters, a repeated measures factorial three-way analysis of variance (ANOVA) is performed with three factors: one between-subjects factor (subject *group*: healthy or AD) and two within-subjects factors (*condition*: eyes open or eyes closed; *electrode locus*). To identify parameters that potentially differentiate between healthy and AD subjects, main effects of the three factors as well as their interaction effects are investigated. The investigation in this step is not intended to localize differences in the parameter to any specific electrode locus. To achieve such a specific localization, a parameter is selected for further investigation only if the main effect of a between-subjects factor (or an interaction effect involving the between-subjects factor) is significant (significance level $\alpha = 0.05$).

Second, the efficacy of the *global* complexity and *global* chaoticity computed from the EEG and EEG sub-bands are investigated using a one-way ANOVA for distinguishing between the two EEG groups. In this step, the global complexity and chaoticity are estimated by averaging the values of the parameters selected in the first step across all 19 loci. It should be noted that this definition of global complexity is different from the one from the literature discussed in Chapter 11 which takes into consideration the spatial evolution of the EEG across loci. In the third step, the *local* complexity and chaoticity in various brain regions are investigated to discover spatial patterns that could not be obtained from the global parameters. To achieve this goal, the one-way ANOVA is performed separately for each locus based on the parameters selected in the first step. In the remainder of this chapter, the parameter name will be preceded by the word *global* or *local* to indicate the source of the parameter. The electrode locus (Fig. 12.1) is used instead of the word *local* wherever specific local parameters are discussed. It is pointed out that the temporal aspect of the spatio-temporal analysis is implicit in the wavelet-chaos methodology and not a separate statistical investigation.

12.3 Results

12.3.1 Complexity and Chaoticity of the EEG: Results of the Three-Way Factorial ANOVA

In the literature, only the complexity and chaoticity of the entire EEG represented by the parameters CD and LLE, respectively, have been investigated and reported to be reduced in AD. This assertion could not be corroborated in this research. The three-way factorial ANOVA revealed no significant differences ($\alpha = 0.05$; $p < 0.05$) in CD and LLE for the AD subjects compared with the healthy subjects regardless of the two within-subject factors - *condition* and *electrode loci* (i.e., no significant interaction effects were reported). Therefore, CD and LLE were not investigated further for specific loci or conditions.

The three-way factorial ANOVA revealed significant differences in θCD, θLLE, and δLLE between the two groups (main effects $p < 0.05$). No significant interaction effects were observed in θCD, implying that the differences in the two groups were present regardless of the within-subject factors (*condition* and *electrode loci*). Significant *group* × *condition* × *electrode locus* interaction effects (in addition to the aforementioned main effects) were observed in θLLE and δLLE, implying that in addition to the primary differences, the changes in spatial patterns (across electrode loci) between the eyes open and eyes closed conditions may be different in the two groups, healthy and AD subjects. Significant *group* × *condition* interaction effects were observed in αCD and αLLE (although no main effects were observed), implying that although the parameters were not different in the two groups (healthy and AD subjects), the *change* in the parameters between the eyes open and eyes closed conditions may be different in the two groups. In addition, significant *group* × *electrode locus* interaction effects were observed in αLLE, implying the possibility of altered spatial distributions (across electrode loci) of the parameter in the two groups. These observations warranted further investigation of the five parameters θCD, θLLE, δLLE, αCD, and αLLE in order to localize the changes to specific electrode loci and the eyes open or eyes closed condition.

12.3.2   Global Complexity and Chaoticity

The efficacy of global complexity and chaoticity for discriminating between the two groups is investigated for both conditions: eyes open and eyes closed individually using one-way ANOVA. For the eyes open condition, no significant differences were observed in the *global* values of θCD, θLLE, δLLE, αCD, and αLLE. For the eyes closed condition, the *global* θLLE and αLLE were found to be significantly reduced, which could account for the significant *group* × *condition* interaction effects observed from the three-way factorial ANOVA. A surprising finding was a significant increase in the *global* αCD, which will be discussed shortly in this chapter.

12.3.3   Local Complexity and Chaoticity

To discover differences between the two groups of subjects with respect to changes in spatial patterns of complexity and chaoticity in the eyes open and eyes closed conditions individual one-way ANOVAs were performed for each electrode locus under the two conditions. In the eyes open condition, δLLE and θLLE computed from a specific electrode locus in the right frontal area (F_4) were significantly reduced in AD subjects. In the eyes closed condition, αLLE was significantly reduced in AD patients in the frontal midline (F_Z) and left occipital (O_1) areas. θLLE was significantly reduced in the right frontal (FP_2) and left parietal (P_7) areas and δLLE was significantly reduced in the left parietal (P_3) area. αCD was significantly increased at the C_4 locus (right central area). These findings could account for the significant *group × condition, group × electrode locus*, and *group × condition × electrode locus* interactions observed from the three-way factorial ANOVA.

In Figs. 12.3 and 12.4, all loci where local parameters show significant differences are shaded in gray. The light and dark shades of gray will be explained in the following section. Overall, all changes are localized to the right frontal and left parieto-occipital regions.

12.4   Discussion

12.4.1   Chaoticity versus Complexity

The distribution of significant parameters obtained from both eyes closed and eyes open conditions is shown in Fig. 12.3, with the light gray circles representing chaoticity and the dark gray circles representing complexity. In general, the LLE appears to be much more consistent in distinguishing AD patients from healthy control subjects, which appears to imply that the EEG chaoticity is reduced in AD subjects more consistently than EEG complexity. This phenomenon may not have been discovered previously due to two possible reasons. One, the method of computation of LLE used in our methodology

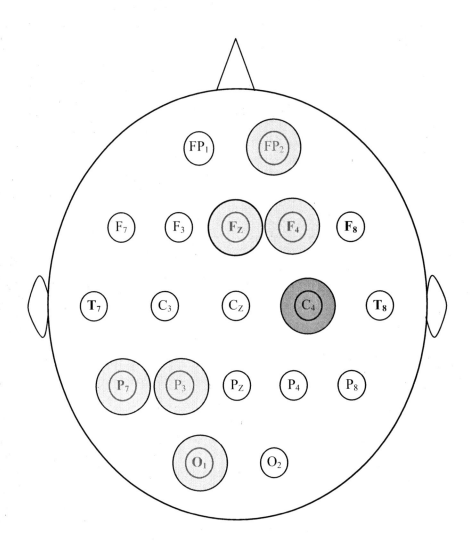

FIGURE 12.3
Electrode loci showing the relative distribution of statistically significant parameters for chaoticity (light gray circles) and complexity (dark gray circle) obtained from both eyes closed and eyes open conditions

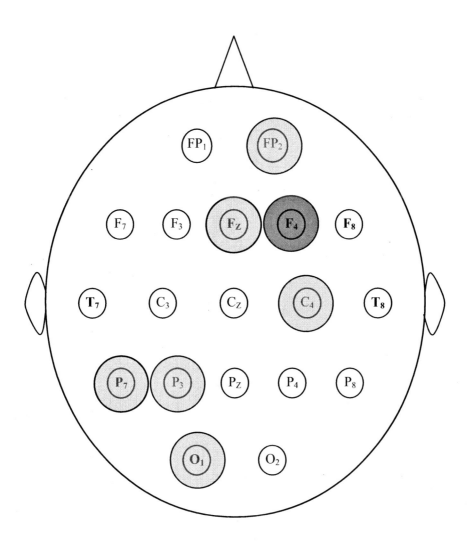

FIGURE 12.4
Electrode loci showing the relative distribution of statistically significant parameters (both chaoticity and complexity) in the eyes closed condition (light gray circles) and eyes open condition (dark gray circles)

is particularly suited to the characterization of non-linear dynamics of non-stationary EEG-like signals (Iasemidis et al., 2000a; Petrosian et al., 2001; Adeli et al., 2003, 2007; Ghosh-Dastidar and Adeli, 2007; Ghosh-Dastidar et al., 2007, 2008). Two, the issue has not been studied previously from the perspective of EEG sub-bands.

12.4.2   Eyes Open versus Eyes Closed

The wavelet-chaos methodology appears to be more effective for extracting meaningful markers of abnormality from the eyes closed condition than the eyes open condition. The relative distribution of significant parameters (chaoticity and complexity) is shown in Fig. 12.4, with the light gray circles representing the eyes closed condition and the dark gray circles representing the eyes open condition.

From previous research (Pritchard et al., 1991), where only EEG complexity was studied, there appeared to be some evidence that the differences between the two groups are more significant in the eyes open condition than in the eyes closed condition. In the earlier research, no significant differences were found in the eyes closed condition possibly because 1) individual sub-bands had not been investigated with respect to their underlying chaotic dynamics and 2) the chaoticity was not studied as a marker. In our research, the opposite was found to be true. For the eyes open condition, the CD appears to be of little use for differentiating AD patients from healthy controls irrespective of whether the global or local CD is employed. Another possible explanation for this apparent contradiction may be that the conclusions of Pritchard et al. (1991) are based on *dimensional complexity* which, although similar to the CD, is not computed by the Takens method employed in our methodology.

The availability of markers of abnormality obtained from eyes closed EEG has two advantages. First, there is no need for the patients to keep their eyes open and gaze steady during EEG recording to avoid eye blink and ocular artifacts. This is of special interest for the care of AD patients who find it very difficult to maintain such steady conditions. As a result, the process is quite uncomfortable for the patient and the EEG is nevertheless characterized by excessive artifacts. Second, due to the excessive artifacts in the eyes open EEG, discrete artifact-free segments of the EEG often have to be patched together to obtain EEGs of the desired duration. This requires significant offline processing of the EEG. Moreover, the effects of the patching process on the subsequent chaos analysis have not been studied in detail. These effects, however, may be significant because of the implicit mismatch in characterizing *continuous* brain dynamics by *discontinuous* EEGs.

Why is the eyes closed EEG more effective than the eyes open EEG for distinguishing AD patients from healthy control subjects? We provide two possible explanations for this. One, the wavelet-chaos methodology and the sub-band analysis accurately characterize the nonlinear dynamics of non-stationary EEG-like signals with respect to the EEG chaoticity. As a result, new potential markers of abnormality were discovered. Two, the neurological processes, especially those governing the observed decrease in EEG chaoticity, in the eyes closed condition lead to differences between AD patients and healthy control subjects. The eyes closed condition represents the internal brain dynamics without the modulation associated with visual attention and the resultant cognitive processing in the eyes open condition. It is possible that exposure to external stimuli raises the chaoticity of the brain in AD to (or close to) the level of a healthy brain of approximately the same age, thus making the two groups indistinguishable on this basis.

12.5   Concluding Remarks

Since the EEG is an overall representation of brain dynamics, it opens up the possibility that the observed changes in the parameters quantifying chaos in the EEG are actually the result of the superimposition of multiple processes underlying the EEG. In this chapter, these underlying processes are investigated using the component physiological sub-bands of the EEG which can be assumed to represent these processes at a finer level.

It is found that when the statistical analysis is based on the entire EEG, the LLE or the CD cannot be used as a discriminating parameter between the two groups. However, when the statistical analysis is performed on the EEG sub-bands, it is observed that the CD as well as the LLE from certain physiological sub-bands and loci may be employed to distinguish between the groups. As a result of this investigation, it is concluded that changes in the dynamics are not spread out equally across the spectrum of the EEG, but instead are limited to certain frequency bands. Moreover, the changes are not globally spread over the entire brain but localized to specific electrode loci.

Eleven potential markers of abnormality were discovered using the wavelet-chaos methodology, 2 in the eyes open condition (F_4 δLLE and θLLE) and 9 in the eyes closed condition (*global* θLLE, αCD, and αLLE; and FP_2 θLLE, P_7 θLLE, P_3 δLLE, C_4 αCD, F_Z αLLE, and O_1 αLLE). Other markers such as the neural complexity in the eyes open condition dubbed *dynamic responsivity* have been reported in the literature (Pritchard et al., 1991). These markers may well represent different aspects of AD and can be used to complement each other in clinical applications. The availability of multiple potential discriminating parameters will result in increased accuracy of EEG classification for AD patients which could form the basis for automated diagnosis of AD in a clinical setting.

Part IV

Third Generation Neural Networks: Spiking Neural Networks

13

Spiking Neural Networks: Spiking Neurons
and Learning Algorithms

13.1   Introduction

Artificial neural networks (ANNs) are simplified mathematical approximations
of biological neural networks in terms of structure as well as function. In
general, there are two aspects of ANN functioning: (1) the mechanism of
information flow starting from the presynaptic neuron to the postsynaptic
neuron across the network and (2) the mechanism of learning that dictates
the adjustment of measures of synaptic strength to minimize a selected cost
or error function (a measure of the difference between the ANN output and
the desired output). Research in these areas has resulted in a wide variety
of powerful ANNs based on novel formulations of the input space, neuron,
type and number of synaptic connections, direction of information flow in the
ANN, cost or error function, learning mechanism, output space, and various
combinations of these.

Ever since the conception of the McCulloch-Pitt neuron in the early 1940s
and the perceptron in the late 1950s (Adeli and Hung, 1995), ANNs have been
evolving toward more powerful models. Advancement in the understanding of
biological networks and their modes of information processing has led to the

development of networks such as feedforward neural networks (Rumelhart et al., 1986), counter-propagation neural networks (Grossberg, 1982; Hecht-Nielsen, 1988; Adeli and Park, 1995a; Sirca and Adeli, 2001; Dharia and Adeli, 2003), radial basis function neural networks (Karim and Adeli, 2002a, 2003; Liu et al., 2007; Mayorga and Carrera, 2007; Pedrycz et al., 2008), recurrent networks (Hopfield, 1982; Zhang et al., 2007; Schaefer and Zimmermann, 2007; Panakkat and Adeli, 2007), self-organizing maps (Kohonen, 1982; Carpenter and Grossberg, 1987), modular neural networks, fuzzy neural networks (Adeli and Karim, 2000; Adeli and Jiang, 2003, 2006; Sabourin et al., 2007; Rigatos, 2008; Jiang and Adeli, 2008a), and spiking neural networks (Sejnowski, 1986; Maass, 1996, 1997b; Iglesias and Villa, 2008; Grossberg and Versace, 2008).

Feedforward ANNs are the most common and utilize various mechanisms for a forward transfer of information across the neural network starting from the input node to the output node. The popularity of feedforward ANNs stems from their conceptual simplicity and the fact that the primary (but not the only) mode of information transfer in both real and artificial neural networks is feedforward in nature (Adeli and Hung, 1994; Adeli and Park, 1995b; Adeli and Karim, 1997; Adeli and Jiang, 2003). In fact, other modes of information transfer often involve or are based on feedforward mechanisms to some degree.

Although ANNs have gone through various stages of evolution, until recently, there had not been many attempts to categorize generations of neural networks (Maass, 1997b). This is a particularly difficult task because ANN developments have branched out in many directions and it would not be accurate to label one advancement as more significant than another. In addition, such a categorization is subjective and dependent on what is considered advancement. However, following Maass (1997b), if a single clearly identifiable, major conceptual advancement were to be isolated, it would be the development of the mathematically defined activation or transfer function as the information processing mechanism of the artificial neuron (Ghosh-Dastidar and Adeli, 2009b).

13.2 Information Encoding and Evolution of Spiking Neurons

Studies of the cortical pyramidal neurons have shown that the timing of individual spikes as a mode of encoding information is very important in many biological neural networks (Sejnowski, 1986; Maass, 1996, 1997b). Biologically, a presynaptic neuron communicates with a postsynaptic neuron via trains of spikes or action potentials. Biological spikes have a fixed morphology and amplitude (Bose and Liang, 1996). The transmitted information is usually encoded in the frequency of spiking (*rate encoding*) and/or in the timing of the spikes (*pulse encoding*). Fig. 13.1 shows biological synapses connecting a presynaptic neuron to a postsynaptic neuron.

Pulse encoding is more powerful than rate encoding in terms of the wide range of information that may be encoded by the same number of neurons (Maass, 1997c). In fact, rate encoding can be considered to be a special (and less powerful) case of pulse encoding because in pulse encoding the spike timings are known, and the average firing rate can be easily computed based on that information. However, in rate encoding the ability to encode complex spike trains is reduced significantly because the temporal information about individual spikes is lost.

The early *first generation* neurons developed in the 1940s and 1950s did not involve any encoding of the temporal aspect of information processing. These neurons acted as simple integrate-and-fire units which fired *if* the *internal state* (defined as the weighted sum of inputs to each neuron) reached

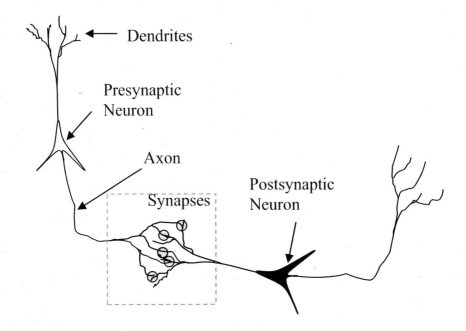

FIGURE 13.1
Biological synapse connecting a presynaptic neuron to a postsynaptic neuron

a threshold. It did not matter when the threshold was exceeded. Translating this assumption to a biological perspective, it implied that all inputs to the neuron were synchronous, i.e., contributed to the internal state at exactly the same time and therefore could be directly summed. However, unlike biological neurons, the magnitude of the input was allowed to contribute to the internal state. Arguably, this may have represented a primitive form of rate encoding in the sense that a larger input (representing a higher firing rate of the presynaptic neuron) may cause the postsynaptic neuron to reach the threshold. For the sake of simplicity, the mathematical abstraction avoided the modeling of the actual spike train and the input from the presynaptic neuron approximated the average firing rate of the presynaptic neuron. The *fire* state for the postsynaptic neuron was a binary-valued output which returned a value of 1 if the neuron fired and 0 otherwise. This implied that the output from the postsynaptic neuron was not based on rate encoding.

The *second generation* neurons developed from the 1950s to 1990s were also based loosely on rate encoding and defined the internal state in a similar manner. However, they used a mathematically defined activation function, often a smooth sigmoid or radial basis function (RBF), instead of a fixed threshold value, for output determination (Maass, 1996). In the postsynaptic neuron, the activation function was used to transform the input into a proportionate output which approximated the average firing rate of the postsynaptic neuron. With this development, it became possible for the output to be real-valued. In contrast to the first generation neurons, in this case even the postsynaptic neuron could generate rate encoded information. This model gained widespread acceptance as processing elements in feedforward ANNs. This popularity was further increased due to Rumelhart's backpropagation (BP) learning algorithm (Rumelhart et al., 1986) developed for these ANNs that enabled supervised learning. Since the BP algorithm was constrained by its requirement of a continuous and differentiable activation function, a significant portion of the ensuing research focused on finding more appropriate continuous and differentiable activation functions. This model was significantly more powerful than the one based on first generation neurons and could solve complex pattern recognition problems (the most notable early example was the XOR problem) (Hung and Adeli, 1993; Park and Adeli, 1997; Adeli and Wu, 1998; Adeli and Samant, 2000; Sirca and Adeli, 2001, 2003; Dharia and Adeli, 2003; Panakkat and Adeli, 2007; Jiang and Adeli, 2008b). However, the computational power of the neuron still did not reach its full potential because the temporal information about individual spikes was not represented.

In the past decade or so, to overcome this shortcoming, neurons that can communicate via the precise timing of spikes or a sequence of spikes have been developed and adapted for ANNs. These neurons have been dubbed *spiking neurons*. In the literature, these spiking neurons have been referred to as *third generation* neurons (Maass, 1997b). Similar to the first generation neurons, spiking neurons act as integrate-and-fire units and have an *all or none* response. The spiking neuron, however, has an inherent dynamic nature characterized by an internal state which changes with time. Each postsynaptic neuron fires an action potential or spike at the time instance its internal state exceeds the neuron threshold. Similar to biological neurons, the magnitude of the spikes (input or output) contains no information. Rather, all information is encoded in the timing of the spikes (i.e., pulse encoding), as discussed in the next section.

13.3   Mechanism of Spike Generation in Biological Neurons

In general, action potentials or spikes from various presynaptic neurons reach a postsynaptic neuron at various times and induce *postsynaptic potentials* (PSPs). The PSP represents the internal state of the postsynaptic neuron in response to the presynaptic spike. Figure 13.2 shows an action potential from the presynaptic neuron and the resulting PSP induced in the postsynaptic neuron. Figure 13.2 shows the changes in the PSP based on the characteristics of the synapse such as travel time or delay through the synapse, strength of the synaptic connection, and other biological factors some of which are still not completely understood. Multiple neurons, each with multiple spikes, induce multiple PSPs over time. The PSPs are temporally integrated to compute the internal state of the postsynaptic neuron over time. The postsynaptic neuron

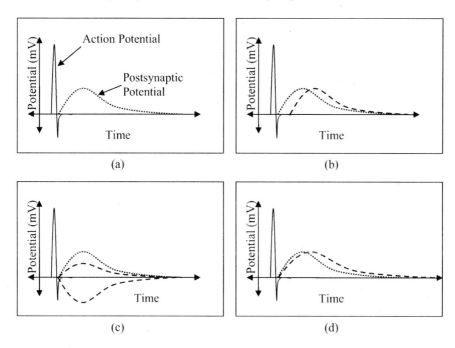

FIGURE 13.2
Spiking neuron response showing the action potential or spike (solid line) and the postsynaptic potential (PSP) in (a) typical case (dotted line), (b) delayed response, (c) weighted response (negative weighted response is inhibitory PSP), and (d) stretched response. All modified responses are shown with a dashed line.

fires a spike when the integrated internal state crosses a threshold (Figs. 13.3 and 13.4). A spike train consists of a sequence of such spikes.

The effects of various presynaptic spike trains on the postsynaptic potential and the postsynaptic output spike train are illustrated in Fig. 13.5. In the first two cases, Fig. 13.5(a) and (b), each spike train is considered individually whereas in the third case, Fig. 13.5(c), the combined effect of the two spike trains shown is illustrated. Each spike train consists of a sequence of three spikes. The first and the third spikes in the presynaptic spike trains occur at the same time instant. The timing of the second spike, however, is different in the two cases. From the perspective of rate encoding, both these spike trains

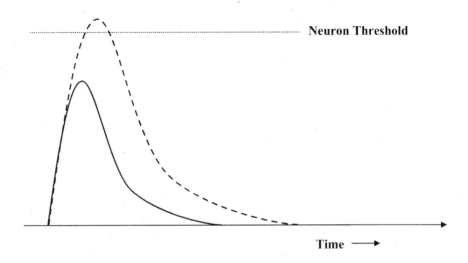

FIGURE 13.3
A larger PSP enables the neuron to reach the neuronal threshold

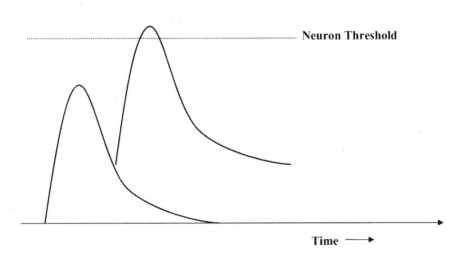

FIGURE 13.4
Multiple superimposed PSPs also enable the neuron to reach the neuronal threshold

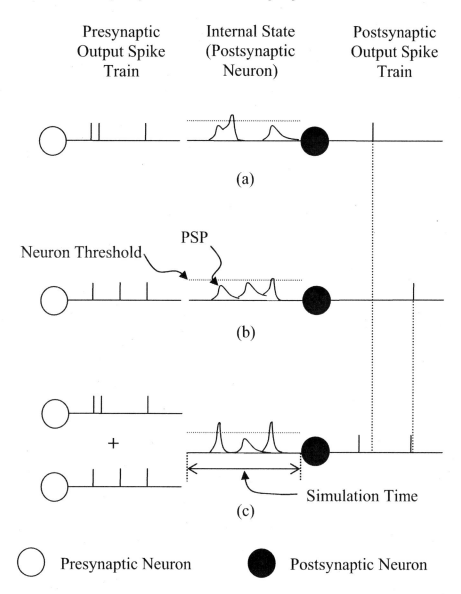

FIGURE 13.5
The effect of various presynaptic spike trains on the postsynaptic potential and the postsynaptic output spike train. (a) and (b) show two spike trains and their individual effects on the postsynaptic neuron, and (c) shows the combined effect of the aforementioned two spike trains on the postsynaptic neuron.

are identical, i.e., the average firing frequency is identical (three per given time period). This highlights the approximate nature and lower computational power of rate encoding which makes it impossible to differentiate between the two cases in Fig. 13.5(a) and (b).

In contrast, the timing of the spikes is considered in pulse encoding. Each spike in the spike train induces a PSP at the time instant it reaches the postsynaptic neuron. The PSPs are temporally integrated to compute the internal state of the postsynaptic neuron over time, as shown in Fig. 13.5. The internal states in the two cases are entirely different and their values exceed the neuronal threshold at different times. This leads to different output spike times from the postsynaptic neuron. An additional source of variation in the PSP is the dependence of the internal state of the postsynaptic neuron on the time of its own output spike. The internal state of a postsynaptic neuron in response to a presynaptic spike is shown in Fig. 13.6. Had the threshold not been exceeded the internal state of the neuron in Fig. 13.6 would have been represented by the dashed line. The solid line in Fig. 13.6 shows the internal state of the neuron when the threshold is exceeded. Immediately after the firing of an output spike, the internal state of the neuron exhibits a sharp decrease as a result of various biological processes. This phase is known as *repolarization* (Fig. 13.6) (Bose and Liang, 1996; Kandel et al., 2000).

In the third case shown in Fig. 13.5(c), both presynaptic spike trains are input simultaneously to the postsynaptic neuron by two presynaptic neurons. In this case, the internal state of the postsynaptic neuron is not simply the sum of the internal states in the first two cases. An additional factor needs to be considered for the postsynaptic neuron. After the firing of a spike and the resultant sharp decrease in the internal state of the neuron, the internal state is kept at a value lower than the resting potential of the neuron (Fig. 13.6) by various biological processes that are beyond the scope of this discussion.

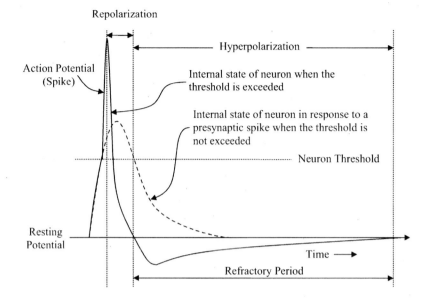

FIGURE 13.6

The internal state of a postsynaptic neuron in response to a presynaptic spike (not shown in the figure) showing the action potential, and repolarization and hyperpolarization phases.

This phase is known as *hyperpolarization* and is shown in Fig. 13.6 (Bose and Liang, 1996; Kandel et al., 2000). As a result, it becomes difficult for the neuron to reach the threshold and fire again for a certain period of time, known as *refractory period* (Fig. 13.6). The internal state of the postsynaptic neuron is obtained by the algebraic summation of the internal states in the first two cases and modified during the repolarization and hyperpolarization phases. The three processes of summation, repolarization, and hyperpolarization lead to the postsynaptic neuron firing output spikes at times different from those for the first two cases. In Fig. 13.5, the first spike in the third case occurs earlier than the first spike in the first case because the postsynaptic neuron in the third case exceeds the threshold value earlier. The three cases shown in Fig. 13.5 highlight the importance of the timing of spikes in the presynaptic spike train for encoding information.

13.4   Models of Spiking Neurons

Spiking neurons can be modeled in many different ways. A number of detailed mathematical or *biophysical* models have been developed to quantitatively characterize neuronal behavior based on detailed modeling of the neuronal membrane potential and ion channel conductances (Hodgkin and Huxley, 1952; Rinzel and Ermentrout, 1989; Hille, 1992; Ermentrout, 1996; Hoppensteadt and Izhikevich, 1997; Izhikevich, 2003). Networks of such neuronal models have proved to be very valuable in studying the behavior of biological neural networks, neuronal learning mechanisms such as long-term potentiation and depotentiation, and neurotransmitter-based signaling (Izhikevich, 2007). Some such models have been discussed in Chapter 10 in the context of Alzheimer's disease. Izhikevich et al. (2004) employed a large network of spiking neuron models described in Izhikevich (2003) with known firing patterns in the cerebral cortex to study self organization in such a network. Recently, Iglesias and Villa (2008) also investigated a large network of spiking neurons described by a different neuron model and various biological processes, and observed a similar pattern of self organization and preferential firing patterns in the neural network.

Such detailed networks are important for studying the effects of various spiking patterns especially in terms of network activation patterns, i.e., which neurons are activated and whether certain neurons are activated more often than others. The level of detail in such models, although ideal for reproducing electrophysiological responses accurately, increases the complexity of the models, makes them difficult to analyze, and imposes a significant computational burden (Abbott and Kepler, 1990; Kepler et al., 1992; Ghosh-Dastidar and Adeli, 2007). As a result, these networks have not been used for real-world classification or pattern recognition tasks that have typically been the domain of traditional ANNs. It may be argued that real-world data can be encoded as spike trains and classified using such biologically plausible networks. However, from the studies of self organization in such networks (Izhikevich et al., 2004; Iglesias and Villa, 2008), it appears that there may be a biological preference for certain spike patterns. In the opinion of the authors, appropriate biologically plausible spike train encoding methodologies should be investigated thoroughly for use with such networks.

The approach to designing such biophysical neuronal models is a *bottom-up* approach. Various detailed neuronal characteristics such as properties of the cell membrane, ion channels, mathematical formulations of learning such as long term potentiation and depotentiation, Hebbian learning, and spike time dependent plasticity (STDP) are modeled separately. These separate models are integrated to obtain the overall characterization of neuronal dynamics. On the other hand, phenomenological models represent a *top-down* approach in which the overall behavior of the individual neuron is modeled mathematically. The precise details of neuronal behavior at the level of ion channels or neurotransmitter molecules are not modeled explicitly. Sometimes, the details are approximately derived but only as secondary phenomena. Spike response models are examples of such phenomenological models that are simpler than the detailed biophysical models and offer a compromise between computational burden and electrophysiological detail (Ermentrout and Kopell, 1986; Rinzel and Ermentrout, 1989; Gerstner, 1995; Kistler et al., 1997; Izhikevich, 2001; Gerstner and Kistler, 2002). Therefore, spike response models are preferred for systemic studies of memory, neural coding, and network dynamics.

13.5   Spiking Neural Networks (SNNs)

It is interesting to note that the phenomenological models can be further simplified to lesser and lesser realistic models until a point is reached where even the spikes are not modeled, which leads to the first and second generation neurons. Evidently, the computational burden reduces significantly and so does the degree of biological realism. In some ways, the level of modeling detail is a function of the available computing power. For instance, consider the ANNs based on second generation neurons. For decades, the modeling of the neurons was limited by the available computing power because the hardware was unable to support large ANNs based on detailed neuronal models. This limitation dictated the design of the learning algorithms. Subsequently, even when advances were made in computing power, proportionate advances were not made in the complexity of the neuronal models because the existing learning algorithms were not compatible with the detailed models.

As a result, two distinct research areas emerged. The field of *artificial neural networks* focused on the behavior of large networks of neuron-like processing units (i.e., the second generation neurons), which were primitive and oversimplified formulations of biological neurons. However, it was demonstrated that even such networks were capable of learning using pseudo-realistic learning algorithms such as backpropagation. ANNs were applied with great success to pattern recognition, classification, and completion tasks in a wide variety of areas. The other field became known as *computational neuroscience*. Within this broad interdisciplinary field, the detailed biophysical and phenomenological models were primarily used in relatively smaller networks to study electrophysiological processes, pattern generation, and the dynamic behavior of small groups of neurons. There have also been studies involving very large numbers of interconnected biophysical neuron models. However, it has not been possible to use such networks of detailed neurons in a manner similar to ANNs for large real-world pattern recognition and classification tasks, as mentioned in the beginning of this chapter.

Recent advances and the availability of computing power have increased the overlap between the two fields. On the one hand, the processing units, networks, and learning algorithms for ANNs have become biologically more realistic. On the other hand, networks of biophysical neurons have become increasingly larger in size and the biophysical models more detailed. The available computing power still limits the use of the detailed models in large biophysical neural networks for pattern recognition and classification tasks. As the computing power becomes more readily available, suitable learning algorithms are also being developed for such models. The development of spiking neural networks (SNNs) was the next logical step toward achieving this goal (Ghosh-Dastidar and Adeli, 2009b).

Simply stated, SNNs are networks of spiking neurons. The SNN architecture is normally similar to that of a traditional ANN. The processing unit, however, is a spiking neuron, which is typically modeled by a phenomenological model such as a spike response model. As discussed earlier, the use of the biophysical models in certain applications of SNNs is less common due to the computational burden. Therefore, for the purpose of this book, a distinction is made between SNNs that use phenomenological models and networks that use biophysical models. Only research on the former is reviewed in the rest of this book. Since the primary purpose of the SNNs is to learn, a discussion of learning algorithms is also covered in this book.

13.6   Unsupervised Learning

As discussed earlier in Chapter 5, unsupervised learning is based solely on the characteristics of the data. The network *learns* the patterns in the data without being guided by any external cues regarding a *desired* outcome. One advantage of unsupervised clustering is the lower computational burden because the process eliminates the need for multiple iterations through a training dataset, which is typically required for supervised learning algorithms such as gradient descent and its variants. As a result, it is not surprising that most initial applications of SNNs were restricted to applications of unsupervised learning.

An early SNN model was presented by Hopfield (1995) where the stimuli were represented by the precise timing of spikes and the spike pattern was encoded in the synaptic delays. The neurons acted similar to an RBF neuron, i.e., they fired when the input spike pattern was similar to the pattern encoded as the center of the RBFs. Otherwise, the neuron did not fire. This similarity was modeled by a distance function between the patterns. Similarity or dissimilarity was decided by a fixed threshold. Soon after, an STDP-like learning rule was presented for a similar RBF neuron that used the spike time difference between the presynaptic and postsynaptic spikes as the basis for learning (Gerstner et al., 1996). Using a similar concept, Maass and Natschläger (Maass and Natschläger, 1997, 1998a,b) modeled the temporal encoding of associative memory using a Hopfield network composed of spiking neurons. In their work they used a traditional recurrent network architecture. Using these models, it was shown that unsupervised learning and self-organization were possible in networks of spiking neurons (Natschläger and Ruf, 1998; Natschläger et al., 2001).

These models were further enhanced to analyze spatial and temporal patterns in the input space and cluster the input data. It was demonstrated that the clustering based on these models converged reliably even when the input data were corrupted by noise (Natschläger and Ruf, 1998, 1999). For the sake of simplicity, neurons in these models were restricted to the emission of a single spike. Given this limitation, the connection between two SNN neurons was modeled by multiple synapses (Natschläger and Ruf, 1998) in order to enable the presynaptic neuron to affect the postsynaptic neuron by inducing PSPs of varying magnitudes at various time instants.

Bohte et al. (2002b) employed a network similar to Natschläger and Ruf (1998) and demonstrated that SNNs are capable of clustering real-world data. They used a population encoding scheme to encode the data to improve the accuracy of the SNN. Based on this encoding scheme, they proposed a multi-scale encoding where each dimension of the input space was encoded by multiple neurons with overlapping Gaussian fields. They reported good performance of the network with the Fisher iris dataset and an arbitrary image segmentation application compared to traditional clustering methods such as k-means clustering and self organizing maps. Bohte et al. (2002b) also reported an increased robustness of their model with respect to noise in the data. They extended their model to a multi-layer RBF SNN that performed hierarchical clustering of the data. More recently, Gueorguieva et al. (2006) investigated a similar network architecture with an STDP-based learning mechanism discussed earlier and arrived at the same conclusions regarding efficiency and noise. Following the work of Bohte et al. (2002b), Panuku and Sekhar (2007) present a variation of the learning algorithm in which the weights are adjusted in stages, i.e., the weights between the input and hidden layers are adjusted first and the weights between the hidden and the output layers are adjusted next. It is shown that this algorithm can effectively separate linearly separable as well as interlocking clusters.

13.7   Supervised Learning

Although unsupervised learning was demonstrated in SNNs with a recurrent architecture, until recently, spiking neurons were considered to be incompatible with the error backpropagation required for supervised learning in purely feedforward networks. This incompatibility was due to the lack of a continuous and differentiable activation function that could relate the internal state of the neuron to the output spike times. To demonstrate that BP-based learning is possible in such a network, Bohte et al. (2002a) employed an SNN which comprised spiking neurons based on the spike response model originally presented by Gerstner (1995). The architecture of Bohte et al.'s SNN model was patterned after the one by Natschläger and Ruf (1998) where each connection between a presynaptic and postsynaptic neuron was modeled by multiple synapses and the neurons were restricted to the emission of a single spike.

The learning algorithm presented for SNN by Bohte et al. (2002a), *Spike-Prop*, was developed along the lines of the BP algorithm for traditional neural networks (Rumelhart et al., 1986). In SpikeProp, error backpropagation is made possible by assuming that the value of the internal state of the neuron increases linearly in the infinitesimal time around the instant of neuronal firing. In other words, around the instant of neuronal firing the internal state is approximated by a finely discretized piecewise linear function. The learning and classification capabilities of SpikeProp were investigated by applying it to the well-known XOR problem as well as three benchmark problems: Fisher iris plant classification, Wisconsin breast cancer tumor (malignant or benign) classification, and Landsat satellite spectral image classification (Newman et al., 1998). The SNN architecture, as shown in Fig. 13.7(a) is similar to that of a traditional feedforward ANN. However, unlike feedforward ANNs where two neurons are connected by one synapse only, the connection between two SNN neurons is modeled by multiple synapses, as shown in Fig. 13.7(b) (Natschläger and Ruf, 1998; Bohte et al., 2002a). Each synapse has a weight and a delay associated with it. This means that a presynaptic neuron can affect a postsynaptic neuron by inducing PSPs of varying magnitudes at various time instants. The magnified connection in Fig. 13.7(b) displays the temporal sequence of spikes (short vertical lines) from the presynaptic neuron, the synaptic weights (proportionate to the size of the star shaped units in the center), and the resulting PSPs (proportionate to the size of the waveform).

Subsequently, SNN was used with various learning algorithms such as backpropagation with momentum (Xin and Embrechts, 2001; McKennoch et al., 2006), *QuickProp* (Xin and Embrechts, 2001; McKennoch et al., 2006), resilient propagation (*RProp*) (McKennoch et al., 2006), and Levenberg-Marquardt BP (Silva and Ruano, 2005) to improve network training performance. QuickProp is a faster converging variant of the original BP learning rule (Rumelhart et al., 1986) that searches for the global error minimum by approximating the error surface on the basis of local changes in the gradient and weights (Fahlman, 1988). RProp is also a fast variant of the BP algorithm where the weights are adjusted based on the direction of the gradient rather than the magnitude. This strategy is specially effective or *resilient* when the error surface is highly uneven and the gradient is not an accurate predictor of the learning rate (Riedmiller and Braun, 1993). Compared with SpikeProp, the aforementioned improved algorithms reportedly provide faster convergence by 20-80%. Some preliminary research has also been reported regarding the adjustment of other SNN parameters such as neuron threshold, synaptic delays,

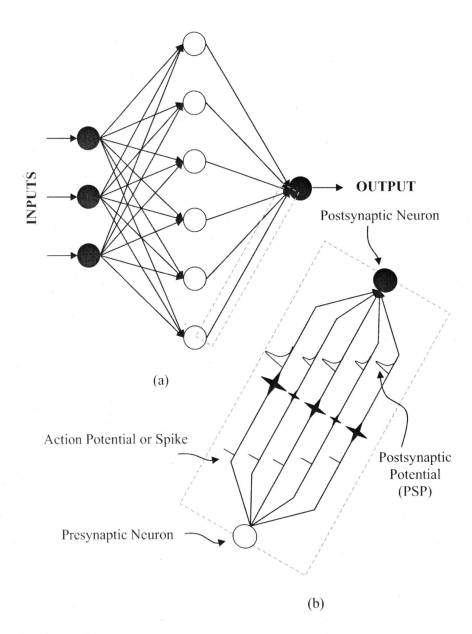

FIGURE 13.7
(a) Spiking neural network architecture; (b) multiple synapses connecting a presynaptic neuron to a postsynaptic neuron

and the time decay constant defining the shape of the PSP (Schrauwen and van Campenhout, 2004).

More recently, McKennoch et al. (2009) presented an alternate model for error backpropagation in an SNN that uses *theta neurons* as the processing units. Theta neurons do not model the actual action potential in time domain, which is a discrete response. Instead, the response is transformed to a phase plane formed by plotting the internal state against the recovery variable (mathematical representation of the recovery state of the inactive ion channels during the refractory period). Finally, the phase plane response is expressed in terms of the phase with respect to time, which is a continuous function (Gutkin and Ermentrout, 1998; Gutkin et al., 2003). Although the theta neuron does not precisely model the postsynaptic spikes, the continuous nature of the phase function makes it an attractive alternative in terms of compatibility with the backpropagation algorithm (McKennoch et al., 2009).

SpikeProp is similar to the BP algorithm where the synaptic weights are adjusted in either batch or incremental processing modes. In incremental processing, the synaptic weights are updated after each training instance is applied. In batch processing, the weights are updated after all the training instances have been applied. In both cases, a pass through all training instances is defined as one epoch. SpikeProp has been applied mostly in the incremental processing mode (Bohte et al., 2002a; Moore, 2002) with one exception (McKennoch et al., 2006). QuickProp and RProp have been applied in the batch processing mode only.

Computationally, SNN training is usually at least two orders of magnitude more intensive than the traditional ANNs for two reasons. First, multiple weights have to be computed for multiple synapses connecting a presynaptic neuron to a postsynaptic neuron. Second, the internal state of each neuron has to be computed for a continuous duration of time, called the *simulation time*, to obtain the output spiking times (Fig. 13.5). The time resolution, called the *time step*, employed for this computation along with the simulation time results in a total of $K \times$ *simulation time/time step* computations per connection between a presynaptic and a postsynaptic neuron (compared to one computation for traditional ANNs). The parameter K is the number of synapses per connection, as shown in Fig. 13.7(b). In addition, the number of convergence epochs is another key factor that affects the actual computation time (real time) required to train the network.

Another difficulty with SNN training is the highly uneven nature of the error surface that can wreak havoc with the gradient descent-based learning algorithms. Slight changes in the synaptic weights result in proportionate changes in the postsynaptic potential. But slight changes in the postsynaptic potential may result in disproportionate changes in the output spike times of the postsynaptic neuron. To overcome this training difficulty various heuristic rules are used to limit the changes of the synaptic weights.

These studies contribute to the current understanding of SNN behavior. However, due to the increased computational complexity and computation time involved in training the SNN for complicated problems with large datasets, the studies extending the original SpikeProp research suffer from three shortcomings. First, extensive parametric studies have not been performed on all algorithms. It is possible that algorithms claimed to be less efficient were so as a result of sub-optimal parameter values. Second, only the number of convergence epochs, and not the actual computation time or the classification accuracy, has been investigated as a performance measure for comparing the learning algorithms. Third, detailed studies are reported only for the XOR problem and small subsets of a benchmark dataset, the Fisher iris dataset.

The primary objective of the authors is to develop an efficient SNN model for epilepsy diagnosis and epileptic seizure detection, a complicated pattern recognition problem where the patterns are unknown and have to be discovered. SNNs represent the next generation of ANNs which model the dynamics of the brain at a greater level of detail. SNN research is still in its infancy and development in this area can potentially lead to more powerful models of learning, and eventually, more accurate and robust classifiers. Toward this objective, three learning algorithms are investigated: SpikeProp (using both incremental and batch processing), and QuickProp and RProp in batch processing mode only. Since the epilepsy diagnosis and epileptic seizure detection problem requires a large training dataset, the efficacy of these algorithms is investigated by first applying them to the XOR and Fisher iris benchmark problems.

Three measures of performance are investigated: number of convergence epochs, computational efficiency, and classification accuracy. For an apple-to-apple comparison of the performance of the learning algorithms a *computational efficiency measure* is defined by dividing the time step by the simulation time and the number of convergence epochs. The larger this number the more efficient the algorithm and the less the required computation time for training. Extensive parametric analysis is performed to identify heuristic rules and optimum parameter values that increase the computational efficiency and classification accuracy. The classification accuracy is evaluated only for the Fisher iris and EEG datasets because they are large enough to be divided into training and testing datasets. The SNN training is performed using randomly selected datasets of various sizes in order to investigate their effect on classification accuracy. Algorithms that consistently perform poorly are discarded. As a result, not all algorithms are tested on all datasets. RProp has not been evaluated on the basis of classification accuracy. Moreover, this research is the first application of SNNs to the EEG classification problem for epilepsy and seizure detection.

13.7.1 Feedforward Stage: Computation of Spike Times and Network Error

In this model, each neuron in the SNN model is limited to the emission of a single spike. Also, the network is assumed to be fully connected, i.e., a neuron in any layer l is connected to all neurons in the preceding layer $l+1$ (layers are numbered backward starting with the output layer, numbered as layer 1). Consequently, a neuron $j\,(\in \{1, 2, .., N_l\})$ in layer l is postsynaptic to N_l+1 presynaptic neurons, where N_l is the number of neurons in layer l. Each presynaptic neuron $i\,(\in \{1, 2, .., N_{l+1}\})$ is connected to the postsynaptic neuron j via K synapses. The number K is constant for any two neurons. The weight of the kth synapse $k\,(\in \{1, 2, .., K\})$ between neurons i and j is denoted by w_{ij}^k. Assuming that presynaptic neuron i fires a spike at time t_i, the kth synapse transmits that spike to the postsynaptic neuron at time $t_i + d^k$ where d^k is the delay associated with the kth synapse. The modeling of synapses is identical for all neurons, and the kth synapse between any two neurons has the same delay, d^k.

The internal state of the postsynaptic neuron j in layer l at time t is expressed as (Bohte et al., 2002a):

$$x_j(t) = \sum_{i=1}^{N_{l+1}} \sum_{k=1}^{K} w_{ij}^k \epsilon(t - t_i - d^k) \tag{13.1}$$

where ϵ represents the spike response function, i.e., the unweighted internal response of the postsynaptic neuron to a single spike. This response can be modeled using a number of different functions. Following Bohte et al. (2002a), in this chapter, the so-called α-function (Gerstner, 1995) is selected as the

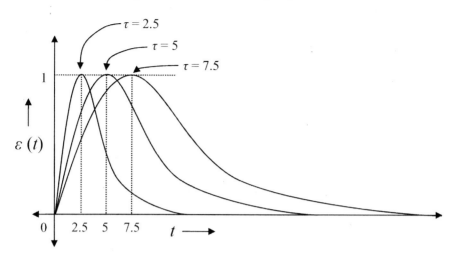

FIGURE 13.8
The spike response function with various time decay constants (τ)

spike response model as follows:

$$\epsilon(t) = \begin{cases} \frac{t}{\tau} e^{1-t/\tau} & \text{when} \quad t > 0 \\ 0 & \text{when} \quad t \leq 0 \end{cases} \tag{13.2}$$

where τ is the time decay constant that determines the spread shape of the function (Fig. 13.8). The postsynaptic neuron fires a spike at the time instant, t_j, when the internal state of the neuron exceeds the neuron threshold, θ. The α-function has a maximum value of 1 at $t = \tau$. Other spike response models (not investigated in the current work) may also be adapted provided that their activation function can be adapted for error backpropagation.

In incremental processing mode, when the neuron j belongs to the output layer ($l = 1$), the output spike time is used to compute the network error as follows (Bohte et al., 2002a):

$$E = \frac{1}{2} \sum_{j=1}^{N_1} (t_j - t_j^d)^2 \tag{13.3}$$

where t_j^d is the desired spike time of the output neuron. In batch processing mode, Eq. (13.3) is replaced by the cumulative error function over all training instances expressed as:

$$E = \frac{1}{2} \sum_{m=1}^{N_T} \sum_{j=1}^{N_1} (t_j - t_j^d)^2 \tag{13.4}$$

where N_T is the number of training instances.

13.7.2   Backpropagation Stage: Learning Algorithms

13.7.2.1   SpikeProp

The SNN is trained by backpropagating the error obtained using Eq. (13.3) or (13.4) and adjusting the synaptic weights such that the network error is minimized. The SNN version of the generalized delta update rule is employed to adjust the synaptic weights. The weight adjustment for the kth synapse between the ith presynaptic and jth postsynaptic neuron is computed as:

$$\Delta w_{ij}^k = -\eta \nabla E_{ij}^k \tag{13.5}$$

where η is the learning rate and ∇E_{ij}^k is the gradient (with respect to the weights) of the error function for the kth synapse between the ith presynaptic and jth postsynaptic neuron. The error gradient at the postsynaptic neuron output spike time instant, $t = t_j$, is computed as:

$$\begin{aligned} \nabla E_{ij}^k &= \frac{\partial E}{\partial w_{ij}^k} \\ &= \frac{\partial E}{\partial t_j} \frac{\partial t_j}{\partial x_j(t_j)} \frac{\partial x_j(t_j)}{\partial w_{ij}^k} \end{aligned} \tag{13.6}$$

Since t_j cannot be expressed as a continuous and differentiable function of $x_j(t_j)$, the term $\partial t_j / \partial x_j(t_j)$ cannot be computed directly. Bohte et al. (2002a) overcome this problem by assuming that $x_j(t_j)$ is a linear function of t_j *around the output spike time instant*, $t = t_j$. Therefore, $\partial t_j/\partial x_j(t_j)$ is approximated numerically as $-1/[\partial x_j(t_j)/\partial t_j)]$. Solving Eq. (13.6), we obtain:

$$\nabla E_{ij}^k = \epsilon(t_j - t_i - d^k)\delta_j \tag{13.7}$$

The term δ_j for a postsynaptic neuron in the output layer is computed as (Bohte et al., 2002a):

$$\delta_j = \frac{-(t_j - t_j^d)}{\displaystyle\sum_{i=1}^{N_2}\sum_{k=1}^{K} w_{ij}^k\epsilon(t_j - t_i - d^k)\left(\dfrac{1}{(t_j - t_i - d^k)} - \dfrac{1}{\tau}\right)} \tag{13.8}$$

For a postsynaptic neuron j in a hidden layer l, the error has to be back-propagated from the output layer for computing the first factor in Eq. (13.6) as:

$$\frac{\partial E}{\partial t_j} = \frac{\partial E}{\partial t_h}\frac{\partial t_h}{\partial x_h(t_h)}\frac{\partial x_h(t_h)}{\partial t_j} \tag{13.9}$$

where the subscript h denotes the hth neuron in layer $l-1$ that is postsynaptic to the postsynaptic neuron j in layer l. As a result, the term δ_j in Eq. (13.7) is computed for a postsynaptic neuron in the hidden layer as (Bohte et al., 2002a):

$$\delta_j = \frac{\displaystyle\sum_{h=1}^{N_{l-1}}\delta_h\sum_{k=1}^{K}\left[w_{jh}^k\epsilon(t_h - t_j - d^k)\left(\dfrac{1}{(t_h - t_j - d^k)} - \dfrac{1}{\tau}\right)\right]}{\displaystyle\sum_{i=1}^{N_{l+1}}\sum_{k=1}^{K} w_{ij}^k\epsilon(t_j - t_i - d^k)\left(\dfrac{1}{(t_j - t_i - d^k)} - \dfrac{1}{\tau}\right)} \tag{13.10}$$

where δ_h is computed for the hth neuron using Eq. (13.8).

To increase the convergence rate, the addition of a momentum term to the delta update rule has been suggested (Xin and Embrechts, 2001) that modulates the weight adjustment on the basis of the weight adjustment in the previous epoch, $\left(\Delta w_{ij}^k\right)_{old}$, as follows:

$$\Delta w_{ij}^k = -\eta\epsilon(t_j - t_i - d^k)\delta_j + \alpha \left(\Delta w_{ij}^k\right)_{old} \qquad (13.11)$$

where α is the momentum factor.

13.7.2.2   QuickProp

The QuickProp weight adjustment is computed as follows (Fahlman, 1988; McKennoch et al., 2006):

$$\Delta w_{ij}^k = \begin{cases} \eta^Q \left(\Delta w_{ij}^k\right)_{old} & \text{when} \quad \left(\Delta w_{ij}^k\right)_{old} \neq 0 \\[2em] -\eta\nabla E_{ij}^k & \text{when} \quad \left(\Delta w_{ij}^k\right)_{old} = 0 \end{cases} \qquad (13.12)$$

where η^Q is the QuickProp learning rate computed adaptively based on the magnitude of the gradient as:

$$\eta^Q = \frac{\nabla E_{ij}^k}{\left(\nabla E_{ij}^k\right)_{old} - \nabla E_{ij}^k} \qquad (13.13)$$

in which $\left(\nabla E_{ij}^k\right)_{old}$ is the gradient with respect to the weights in the previous epoch.

13.7.2.3 RProp

The RProp weight adjustment is computed as follows (Riedmiller and Braun, 1993; McKennoch et al., 2006):

$$\Delta w_{ij}^k = \begin{cases} -\Delta_{ij}^k & \text{when} \quad \nabla E > 0 \\ \\ +\Delta_{ij}^k & \text{when} \quad \nabla E < 0 \\ \\ 0 & \text{when} \quad \nabla E = 0 \end{cases} \tag{13.14}$$

where the term Δ_{ij}^k is defined as:

$$\Delta_{ij}^k = \begin{cases} \eta^+ \left(\Delta_{ij}^k\right)_{old} & \text{when} \quad \nabla E_{ij}^k \cdot \left(\nabla E_{ij}^k\right)_{old} > 0 \\ \\ \eta^- \left(\Delta_{ij}^k\right)_{old} & \text{when} \quad \nabla E_{ij}^k \cdot \left(\nabla E_{ij}^k\right)_{old} < 0 \\ \\ \left(\Delta_{ij}^k\right)_{old} & \text{when} \quad \nabla E_{ij}^k \cdot \left(\nabla E_{ij}^k\right)_{old} = 0 \end{cases} \tag{13.15}$$

in which η^+ and η^- are the RProp learning rates.

14

Improved Spiking Neural Networks with
Application to EEG Classification and
Epilepsy and Seizure Detection

14.1 Network Architecture and Training

14.1.1 Number of Neurons in Each Layer

The SNN architecture consists of an input layer, a hidden layer, and an output layer (Fig. 13.7). The number of neurons in the hidden layer is selected by trial and error. Since the SNN model is based on spike times, inputs to the SNN have to be preprocessed to convert the continuous real-valued input *features* (or classification variables) into discrete spike times. As a result, the number of original features is converted into a new number of features for input to the SNN. This is known as *input encoding*. Similarly, the number of neurons in the output layer depends on the *output encoding* scheme selected for the classification problem. In SNNs the inputs and outputs can be encoded in a variety of ways. This variety, however, is limited by the assumption of only one spike per neuron. The encodings for each one of the three classification problems investigated in this research are discussed separately.

14.1.2   Number of Synapses

The number of synapses, K, between two neurons is an important factor in the SNN architecture. Bohte et al. (2002a) selected a value of 16 for K apparently on a trial and error basis. Subsequent researchers continued using the same number. Values other than 16 were explored to find out their impact on both convergence and the classification accuracy. In general, it was observed that decreasing the number of synapses resulted in reduced convergence rates and classification accuracy whereas an increase in the number of synapses beyond 16 increased the computational effort but did not improve either convergence or classification accuracy. Therefore, in this book a value of $K = 16$ is employed and this issue is not investigated further. Each presynaptic neuron emits one spike and this spike is transmitted through the $K = 16$ synapses sequentially with delays. In the literature, the delays associated with the 16 synapses are modeled by assigning them integer values from 0 to 15 ms (the time unit is *virtual* and is used for modeling purposes only). In this chapter, this range is modified to 1-16 ms because a delay of 0 ms is biologically unrealistic.

14.1.3   Initialization of Weights

SNN training has been reported to be sensitive to the initialization weights which need to be selected carefully. The weight initialization method used in the applications discussed in the book is as follows (Moore, 2002; Ghosh-Dastidar and Adeli, 2007). The neuron threshold for all spiking neurons in the network is selected as $\theta = 1$ and the weight initialization process and heuristic rules (described shortly) are developed around this number (any value can be chosen for threshold as long as the other parameters and heuristic rules are developed around that value). All spike time inputs are set equal to zero and all synaptic weights are set equal to one and the corresponding output spike time is obtained by running the SNN model. Next, the weights of all synapses are selected randomly as real numbers in the range 1-10 and normalized by dividing them by the product of the average of all weights and the output spike time for inputs equal to zero computed earlier. The purpose of this normalization is to limit the range of the initial synaptic weights such that all neurons fire within the simulation time, at least in the first epoch of network training. To ensure the consistency of this method, results are reported in this chapter for ten different sets of weights, dubbed *seeds*, initialized by this method.

14.1.4 Heuristic Rules for SNN Learning Algorithms

In Ghosh-Dastidar and Adeli (2007) extensive parametric analysis was performed to verify heuristic rules proposed in the literature and to identify new heuristic rules with the goal of improving the convergence and classification accuracy and increasing the computational efficiency using the XOR and Fisher iris benchmark problems. These heuristic rules are enumerated as follows:

1. In order to prevent catastrophic changes in the synaptic weights, a lower limit of 0.1 is imposed on the denominator of δ_j in Eqs. (13.8) and (13.10) as suggested by Booij and Nguyen (2005).

2. If at any time during the training of a network, a neuron stops firing, then its contribution to the network error becomes null. During backpropagation of the error, the resulting weight change is very small which may not be sufficient to restart the firing of the neuron even after several epochs. This issue, referred to as the *silent neuron problem*, leads to a reduction of the effective network size to a size possibly insufficient to model the classification problem which ultimately affects convergence (McKennoch et al., 2006). It appears that the QuickProp algorithm is

especially susceptible to this silent neuron problem possibly because the weight change in each epoch is directly linked to that in the previous one. This may be mathematically inferred from Eq. (13.12). In this book, the neuron is set to fire at the maximum internal state value if the threshold is not exceeded during the simulation time. Based on this heuristic, every neuron fires during the simulation time.

3. For QuickProp, the heuristic proposed by Fahlman (1988) for the traditional ANN, that is, defining a *maximum growth factor* (μ) as the upper limit for η^Q, is used. Further, η^Q is multiplied by a factor $0 < \beta \leq 1$ to keep the value of η^Q small (McKennoch et al., 2006).

For RProp, weight restrictions proposed by McKennoch et al. (2006) were investigated. No improvement in convergence was observed and even, in some cases, deterioration in convergence was noted. Therefore, no such weight restrictions have been used in the simulation results presented in the chapter.

SNN performance is affected by a large number of parameters that define the spiking neuron, network architecture, and the learning algorithm such as training size, training data, simulation time, time step, learning algorithm learning rates, and the limiting convergence error. For an effective comparison of the SNN learning algorithms, it is imperative for the selected values of the network parameters to be close to optimal. The selection of such optimum values is achieved through extensive parametric analysis. Due to the excessive computation time required for training SNNs with large datasets, it is not possible to investigate all values of the aforementioned parameters with all three learning algorithms. This is compounded by limitless options regarding SNN architectures. Some parameters have previously been studied to some extent and documented reasonably well in the literature. For other parameters, a parametric analysis is performed in three stages corresponding to the three classification problems: XOR, Fisher iris, and EEG (Ghosh-Dastidar and Adeli, 2007) classification problems. As the analysis progresses from a simple problem to a complex classification problem, SNN architectures, parameter values, and learning algorithms that do not meet the goal of increased computational efficiency and classification accuracy are removed from further consideration.

14.2 XOR Classification Problem

14.2.1 Input and Output Encoding

The exclusive OR (XOR) problem has been used as a common benchmark for the initial testing of different ANN models. The data consist of two binary input features and one binary output. The dataset consists of four training samples. If the two inputs are identical ($\{0,0\}$ or $\{1,1\}$), the output is 0. If the inputs are different ($\{0,1\}$ or $\{1,0\}$), the output is 1. The problem is non-trivial as the two classes are not linearly separable. At the same time, the small size (4×2) of the dataset allows fast training of classification models. The inputs to SNN are discrete spike times. SNN outputs are also spike times. Following Bohte et al. (2002a), an input value of 1 is encoded as an early spike time (0 ms) whereas a value of 0 is encoded as a late spike time (6 ms). Bohte et al. (2002a) select an *encoding interval* (difference between the two spike times) of 6 ms for this problem by trial and error.

Spikes at the input times 0 and 6 ms are transmitted from one layer to the next layer with a minimum delay of 1 ms and a maximum delay of 16 ms introduced by the 16 synapses connecting each presynaptic neuron to a postsynaptic neuron. As a result of the delays, the possible output spike times are in the range from $0 + 2 \times 1 = 2$ ms to $6 + 2 \times 16 = 38$ ms. Bohte et al.

(2002a) select the desired output spike times (t_j) for SNN training toward the middle of this range using a trial and error process. When a small (early) value is selected for the desired output spike time, the delayed input spikes will not have any effect on the output spike computation. When a large (late) value is selected for the desired output spike time the simulation time becomes large which reduces the computational efficiency. Following Bohte et al. (2002a) the output values 1 and 0 are encoded as output spike times 10 ms and 16 ms, respectively.

Bohte et al. (2002a) selected the time decay constant τ for the α-function in Eq. (13.2) by trial and error. A small value of τ results in a narrow-shaped PSP (Fig. 13.8). If the PSP is too narrow, PSPs resulting from presynaptic spikes from various presynaptic neurons delayed by different synapses may not overlap. In such a case, summation of the PSPs does not increase the internal state and, as a result, the postsynaptic neuron does not fire. Too large a value of τ results in very wide PSPs (Fig. 13.8) and the internal state of the postsynaptic neuron takes longer to decrease to the resting potential (Fig. 13.6). The increased width of the PSP leads to too much overlap between subsequent PSPs. As a result, the postsynaptic neuron fires equally early for both small and large values of synaptic weights thus becoming insensitive to changes in synaptic weights during the SNN training. Bohte et al. (2002a) observed that $\tau = 7$ ms, a slightly larger value than the encoding interval, models the XOR classification problem accurately. The same value is used for the XOR problem in this chapter.

14.2.2   SNN Architecture

To fit the model to the data, the number of input and output neurons is selected as 2 and 1, respectively. A problem with spike time encoding is that the SNN treats both inputs {0,0} and {6,6} the same because there is no time

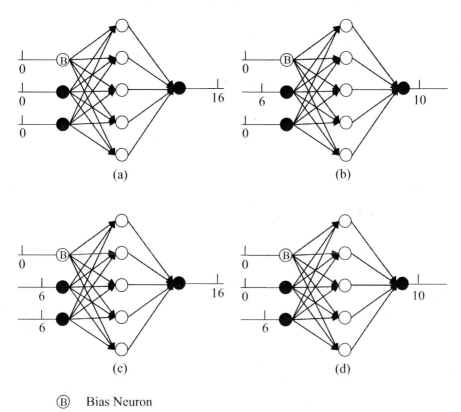

(a)   (b)

(c)   (d)

(B)   Bias Neuron

FIGURE 14.1
SNN architecture for the XOR problem and the input and output encoding
of the four training samples

lag between the two input values. To overcome this problem, a third input
neuron that always fires at 0 ms, dubbed *bias neuron*, is added to the SNN.
As a result the inputs become {0,0,0} and {6,6,0} which are differentiable by
the SNN, as shown in Fig. 14.1. The number of hidden neurons is selected as 5.
This [3 5 1] architecture is selected to be the same as that reported in the SNN
literature in order to compare the results obtained with those published in the
literature and determine the improvements made by the heuristics presented
in this chapter.

14.2.3   Convergence Criteria

Most results in the literature are reported for a limiting convergence mean square error (MSE) value of 0.5 with the exception of Bohte et al. (2002a) who reported a sum squared error (SSE) of 1.0 for the original SpikeProp algorithm within 250 epochs. For both incremental and batch processing learning algorithms, the MSE is equal to the SSE (computed according to Eq. 13.4), averaged over all output neurons and training instances. For the XOR problem, since there are only four training instances and one output neuron, an SSE of 1.0 is equivalent to an MSE of $1.0/(4 \times 1) = 0.25$. In this research, an MSE value of 0.5 is employed and the upper limit for the number of epochs is set to 500. In the literature, non-convergent simulations have been excluded from the statistics and reported separately. That should not be the case. For example, an algorithm that converges in only 50% of the simulations is not necessarily faster than another one that converges slower but in 100% of the simulations. The simulations designated as non-convergent might represent cases of very slow convergence. Therefore, excluding the non-convergent cases yields a bloated estimate of the convergence rate.

14.2.4   Type of Neuron (Excitatory or Inhibitory)

Bohte et al. (2002a), Xin and Embrechts (2001), and others have reported that the SNN training does not converge if any neuron in the hidden layer has a mix of positive and negative weights, as shown in Fig. 14.2. They try to solve this *problem* by designating some neurons to be exclusively excitatory and some others to be exclusively inhibitory, as shown in Fig. 14.2. In this model all synaptic weights of an exclusively excitatory neuron are required to be positive. If the weight adjustment during network training leads to a negative weight, the weight can either be made equal to zero or left unchanged from the previous epoch to meet the requirement of positive weights. Similarly, all synaptic weights of an exclusively inhibitory neuron are required to be negative. If the weight adjustment during network training leads to a positive weight, the weight can either be made equal to zero or left unchanged from the previous epoch to meet the requirement of negative weights. Which scheme (made equal to zero or unchanged) is used in the aforementioned papers is not spelled out. In Ghosh-Dastidar and Adeli (2007) both schemes were investigated and a significant difference in their convergence was noted. The latter scheme results in a faster convergence possibly because a fewer number of weights are changed which prevents drastic changes in the output spike times. Consequently, that is the scheme used in the results presented in the next section whenever exclusively excitatory and inhibitory neurons are used. Recent research indicates the aforementioned assertion regarding the effect of a mix of positive and negative weights may not be correct (McKennoch et al., 2006). This issue has been investigated in this research; the results are presented in the next section.

14.2.5 Convergence Results for a Simulation Time of 50 ms

Table 14.1 summarizes the best SNN convergence results for the XOR problem reported in the literature to date using three different learning algorithms and the corresponding parameter values used. The same simulation time of 50 ms is used in all. A range of values is provided where the exact value has not been reported. The computing times per epoch for the three methods are close to each other. The computational efficiency measure defined earlier is calculated for every algorithm and included in the last column of Table 14.1. Prior research summarized in Table 14.1 indicates that QuickProp and RProp are computationally 4 to 10 times more efficient than SpikeProp. The momentum factor, α in Eq. (13.11), was reported to improve convergence rates by 20% (Xin and Embrechts, 2001; McKennoch et al., 2006) but this was not

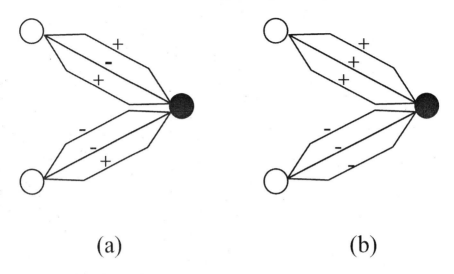

(a)   (b)

FIGURE 14.2
The synapses between two presynaptic neurons and a postsynaptic neuron where (a) no presynaptic neuron is exclusively excitatory or exclusively inhibitory and (b) one presynaptic neuron is designated as exclusively excitatory and the other one as exclusively inhibitory

substantiated by our extensive analysis. This result could not be corroborated in this research. The use of the momentum factor increased the convergence in some cases and decreased the convergence in other cases.

Ghosh-Dastidar and Adeli (2007) investigated the convergence for the three learning algorithms using two different time steps and different learning rates. The results are tabulated in Tables 14.2 and 14.3. A simulation time of 50 ms is used in all cases, the same as that reported in Table 14.1. Table 14.2 summarizes the case where no restrictions are placed on the signs of the weights (no exclusively excitatory or inhibitory neurons are used). Table 14.3 summarizes the results of the case where 4 of the 5 neurons are designated as exclusively excitatory neurons and the fifth neuron is designated as an exclusively inhibitory neuron. Each number in Tables 14.2 and 14.3 is the average of values obtained in ten simulations performed using ten different weight seeds

TABLE 14.1

The best SNN convergence results for the XOR problem reported in the literature to date using three different learning algorithms and a simulation time of 50 ms

	Training Algorithm	Time Step	Learning Rates	Limiting Convergence MSE	No. of Epochs for Convergence	Computational Efficiency
Bohte et al. (2002a)	SpikeProp	0.01	$\eta = 0.01$	0.25	250	0.8×10^{-6}
Moore (2002)	SpikeProp	0.01	$\eta = 0.01$	0.5	121	1.7×10^{-6}
McKennoch et al. (2006)	SpikeProp	0.025	$\eta = 0.05 - 0.50$	0.5	127	3.9×10^{-6}
	QuickProp	0.025	$\eta = 0.05 - 0.20$	0.5	29	1.7×10^{-5}
	RProp	0.025	$\eta^+ = 1.3, \eta^- = 0.5$	0.5	31	1.6×10^{-5}

for network initialization. It should be noted that when all neurons are designated as exclusively excitatory the number of convergence epochs increases drastically. As such, this strategy is not used in SNN. The numbers of epochs for convergence shown in Tables 14.2 and 14.3 in the case of QuickProp and RProp are greater than those reported by McKennoch et al. (2006) because different parameter values are used in this research. Comparable numbers are obtained when the same parameter values are used.

The following three observations are made based on extensive parametric studies performed in this research:

1. *Effect of exclusively excitatory and inhibitory neurons*: The SNN convergence is slower in all cases when neurons are designated to be exclusively excitatory or inhibitory (Table 14.3) compared with the case where mixed signs are allowed for synapses (Table 14.2) with the exception of the SpikeProp algorithm (in both batch and incremental processing modes) only when a very low learning rate of 0.001 is employed.

2. *Effect of time step*: A small time step in the range 0.01-0.025 has been used in the literature for simulation because SNNs could not be trained accurately using larger values. However, the SNN can be trained accurately using larger values of the time step in the range 0.1 and 1 provided that proper heuristic rules (described earlier in the chapter) are incorporated into the learning algorithms and proper parameter values are used. This is an important finding for the computational efficiency of the algorithms. Computational efficiency of all three algorithms increases by a factor of 4 to 40 when a time step of 1 is used compared with a time step of 0.1 assuming all other parameters are kept constant. Using a time step of 1 increases the computational efficiency by a factor of 200-312 compared with a time step of 0.01 used by Bohte et al. (2002a) and

TABLE 14.2
Comparison of SNN computational efficiencies (CE) of three learning algorithms for the XOR problem when no exclusively excitatory or inhibitory neuron is used

Time Step	η	SpikeProp (Batch)		SpikeProp (Incremental)		QuickProp $\beta = 0.1, \mu = 1$		RProp $\eta^+ = 1.2, \eta^- = 0.5$	
		Epoch	CE	Epoch	CE	Epoch	CE	Epoch	CE
1	0.001	180	1.1×10^{-4}	233	8.6×10^{-5}	85	1.1×10^{-4}	93	2.2×10^{-4}
1	0.01	58	3.4×10^{-4}	38	45.3×10^{-4}	62	3.2×10^{-4}	-	-
1	0.1	124	1.6×10^{-4}	500	4.0×10^{-5}	147	1.4×10^{-4}	-	-
0.1	0.001	183	1.1×10^{-5}	252	7.9×10^{-6}	80	2.5×10^{-5}	141	1.4×10^{-5}
0.1	0.01	53	3.8×10^{-5}	46	4.3×10^{-5}	84	2.4×10^{-5}	-	-
0.1	0.1	500	4.0×10^{-6}	500	4.0×10^{-6}	179	1.1×10^{-5}	-	-

TABLE 14.3

Comparison of SNN computational efficiencies (CE) of three learning algorithms for the XOR problem when four exclusively excitatory neurons and one exclusively inhibitory neuron are used

Time Step	η	SpikeProp (Batch)		SpikeProp (Incremental)		QuickProp $\beta = 0.1, \mu = 1$		RProp $\eta^{+} = 1.2, \eta^{-} = 0.5$	
		Epoch	CE	Epoch	CE	Epoch	CE	Epoch	CE
1	0.001	157	1.3×10^{-4}	117	1.7×10^{-4}	200	1.0×10^{-4}	426	4.7×10^{-5}
1	0.01	500	4.0×10^{-5}	500	4.0×10^{-5}	419	4.8×10^{-5}	-	-
1	0.1	500	4.0×10^{-5}	500	4.0×10^{-5}	500	4.0×10^{-5}	-	-
0.1	0.001	143	1.4×10^{-5}	98	2.0×10^{-5}	67	3.0×10^{-5}	194	1.0×10^{-5}
0.1	0.01	443	4.5×10^{-6}	500	4.0×10^{-6}	358	5.6×10^{-6}	-	-
0.1	0.1	500	4.0×10^{-6}	500	4.0×10^{-6}	500	4.0×10^{-6}	-	-

Moore (2002) and by a factor of 14 to 136 compared with a time step of 0.025 used by McKennoch et al. (2006).

3. *Effect of learning rate, η, in SpikeProp and QuickProp*: A learning rate $\eta = 0.01$ results in faster convergence and increases the computational efficiency by a factor of 3 to 10 compared with learning rates of 0.1 and 0.001 (Tables 14.2 and 14.3). This finding is consistent with the literature.

14.2.6 Convergence Results for a Simulation Time of 25 ms

To find out whether the computational efficiency can be further improved by reducing the simulation time, the value of the simulation time is reduced from 50 ms to 20 ms which is expected to be sufficient for temporal integration of spikes considering the last output spike is at 16 ms. The three algorithms are investigated using a time step of 1 ms and a limiting convergence MSE of 0.5. For comparison purposes the learning algorithm learning rates and parameters are kept the same as those in shown in Tables 14.2 and 14.3. Only one learning rate value of $\eta = 0.01$ is employed for SpikeProp and QuickProp.

It is observed that SpikeProp with incremental processing converges in 38 epochs for a simulation time of 20 ms, resulting in a computational efficiency of 1.3×10^{-3} compared with the same 38 epochs for a simulation time of 50 ms, resulting in a computational efficiency of 5.3×10^{-4}. The three batch processing algorithms, SpikeProp with batch processing, QuickProp, and RProp, fail to converge within the maximum set limit of 500 epochs. After several dozen simulation runs and studying the behavior of individual spiking neurons it was discovered that with the smaller simulation time of 20 ms the batch processing algorithms are unable to recover from the silent neuron problem even with the heuristic rule described earlier that forces the neuron to fire. It appears that with a smaller simulation time, the learning algorithms lack the needed time to make the spike time adjustments required to bring the training back on course.

Next, the simulation time for all three algorithms is increased to 25 ms and its effect on the training convergence is studied. The following observations are made:

1. SpikeProp with incremental processing, QuickProp, and RProp converge in 39, 95, and 93 epochs, respectively. The corresponding computational efficiencies are 1.0×10^{-3}, 4.2×10^{-4}, and 4.3×10^{-4}.

2. SpikeProp with batch processing still fails to converge and therefore is removed from the remainder of this investigation. In the rest of the chapter SpikeProp with incremental processing is referred to simply as SpikeProp.

3. For RProp, when learning rates are selected as $\eta^+ = 1.3$ and $\eta^- = 0.8$ as suggested by McKennoch et al. (2006) the number of convergence epochs is reduced to 33 epochs and the computational efficiency is increased from 4.3×10^{-4} to 1.2×10^{-3}.

4. *Optimum QuickProp Parameters*: An extensive parametric analysis was performed to obtain the optimum values of μ and β for convergence using a time step of 1 ms and $\eta = 0.01$. The ranges investigated for the two parameters μ and β are 0.5-4.5 and 0.05-1, respectively. The fastest convergence of 28 epochs (computational efficiency of 1.4×10^{-3}) is achieved when $\mu = 2.5$ and $\beta = 0.1$. The variation of number of convergence epochs with β (using a constant value of $\mu = 2.5$) and with μ (using a constant value of $\beta = 0.1$) are shown in Fig. 14.3 and Fig. 14.4, respectively, as examples. It is observed that not using β, i.e., assuming $\beta = 1$, increases the number of convergence epochs significantly to 366 epochs from the optimum value of 28.

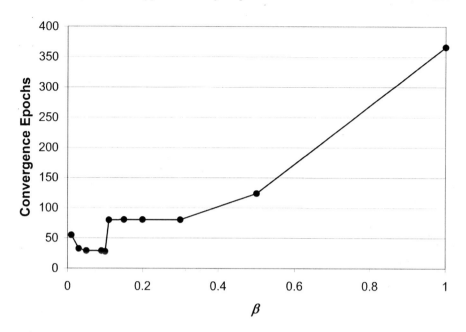

FIGURE 14.3
Number of convergence epochs versus β (using a constant value of $\mu = 2.5$)

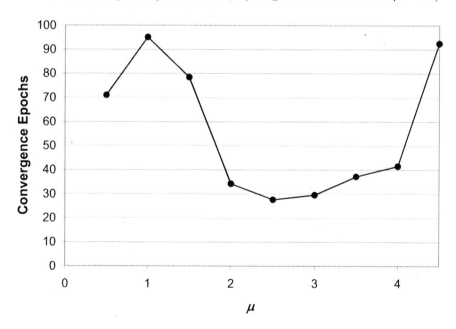

FIGURE 14.4
Number of convergence epochs versus μ (using a constant value of $\beta = 0.1$)

14.2.7   Summary

In this research, the computational efficiencies of the three learning algorithms, SpikeProp, QuickProp, and RProp, for the XOR problem are comparable $(1.0 \times 10^{-3}, 1.4 \times 10^{-3},$ and 1.2×10^{-3}, respectively) when optimum parameter values are employed. This refutes other published works where SpikeProp is reported to be 4-10 times less efficient than QuickProp and RProp (Table 14.1). It is noteworthy that the heuristic rules and optimum parameter values found in this research improve the efficiency of all three algorithms, SpikeProp, QuickProp, and RProp, by factors of 588, 82, and 75, respectively. The increase in computational efficiency as a result of a reduced simulation time and an increased time step becomes even more pronounced for real-world classification problems involving large datasets, to be discussed next.

14.3   Fisher Iris Classification Problem

14.3.1   Input Encoding

The Fisher iris species classification problem consists of four flower features (petal width, petal length, sepal width, sepal length) and three classes (three species of the iris plant: *Versicolor, Virginica,* and *Setosa*) (Fisher, 1936; Newman et al., 1998). The first two classes are not separable linearly. The dataset consists of 150 samples. Unlike the XOR problem, the inputs are continuous and real-valued and therefore cannot be encoded as easily. Following Bohte et al. (2002a), a *population encoding* scheme is used for input encoding. Briefly, each feature is encoded separately by $M > 2$ identically shaped overlapping Gaussian functions centered at M locations. The spread of the Gaussian function is $(1/\gamma)(I_{\max} - I_{\min})/(M - 2)$ where $I_{\max}$ and $I_{\min}$ are the maximum and minimum values for the encoded feature, respectively, and γ is an adjustment factor. The center of the ith Gaussian function is located at $I_{\min} + [(2i - 3)/2][(I_{\max} - I_{\min})/(M - 2)]$ where $i \in \{1, 2, , M\}$. A single input feature value elicits M different responses in the range 0-1. This range of values is converted linearly to the *encoding range* of 0-10 ms (each response value is multiplied by 10) and rounded to the nearest time step selected for the SNN. The input spike times are obtained by subtracting the result from 10 to encode a higher value with an earlier spike time and vice versa (similar to the XOR problem). An example to illustrate this population encoding scheme for converting a real-valued input $I_a = 40$ from a range of real-valued inputs between $I_{\min} = 0$ and $I_{\min} = 100$ using $M = 4$ Gaussian functions is shown in Fig. 14.5. The input ($I_a = 40$) is transformed by the four Gaussian functions into four real numbers 0.001, 0.332, 0.817, and 0.022. These four numbers are multiplied by 10, rounded to the nearest time step (assumed equal to 1 in this example), and the results are subtracted from 10 in order to obtain the discrete SNN input spike times 10, 7, 2, and 10, respectively (Fig. 14.5).

In this research, first a *linear encoding* scheme was investigated where the value for each input feature is converted proportionately to a spike time in the range 0-10 ms and rounded to the nearest time step. However, for Spike-Prop this scheme resulted in classification accuracies approximately 30-50% lower than those obtained using population encoding. As a result, this scheme was not investigated further with any other learning algorithm. Population encoding where real-valued inputs are compressed to a small encoding range of discrete spike times seems to represent the input to SNNs more accurately and is therefore used in this book.

Due to the compression and discretization, two different input feature values close to each other may yield the same spike time response and, therefore, may not be differentiable in the encoding process. This is not desirable when

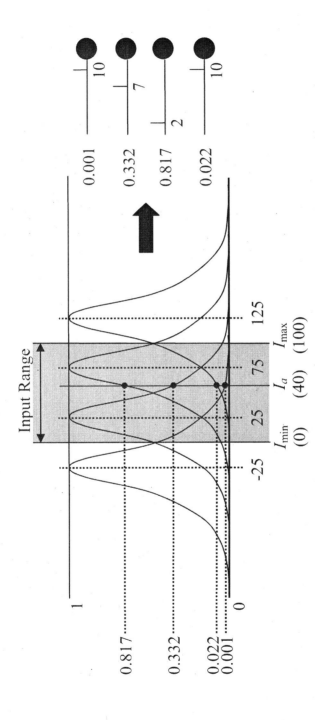

FIGURE 14.5
Population encoding scheme to convert a real-valued input $I_a = 40$ (range of real-valued inputs is from $I_{min} = 0$ to $I_{min} = 100$) to $M = 4$ discrete input spike times $\{10, 7, 2, 10\}$

the classes are not linearly separable. Population encoding transforms each input feature value into M different spike times (instead of 1) which proportionately increases the differentiability. As M increases, the number of SNN input neurons and synaptic weights also increase. Therefore, a trade-off must be made between the computational effort and differentiability.

Several different values of the input encoding adjustment parameter, γ, in the range of 0.5 to 3.0 were investigated in this research. A value of $\gamma = 1.5$ yielded the best classification accuracies which corroborates the observations made by Bohte et al. (2002a). Therefore, the same value of $\gamma = 1.5$ is used in the simulations reported in the remainder of this chapter. The time decay constant τ is selected slightly larger than the encoding interval (i.e., 10 ms) as 11 ms.

14.3.2 Output Encoding

Two output encoding schemes were investigated in this research. In the first scheme, the three classes are separately encoded by three output neurons, which is the standard scheme also used by Bohte et al. (2002a) and Xin and Embrechts (2001). The second scheme employs only one output neuron (McKennoch et al., 2006) and the three classes are represented by the three spike times: 15, 20, and 25 ms. Similar to the XOR problem, these values are selected by trial and error toward the middle of the range of output spike times, $0 + 2 \times 1 = 2$ ms to $10 + 2 \times 16 = 42$ ms. The actual SNN output spike time is assigned to the class that has the closest representative spike time. In this research it was observed that the latter scheme yields higher classification accuracies for RProp and QuickProp by 8-10%. For SpikeProp, however, both schemes yielded similar results. The latter scheme uses a lower number of synaptic weights and, therefore, is employed in the remainder of this chapter.

14.3.3 SNN Architecture

As a result of population encoding, M input neurons are required per input feature plus a bias neuron, resulting in a total of $4M + 1$ input neurons. The purpose of the bias neuron is discussed in Section 14.2.2. Two configurations are investigated in this research. The first configuration employs 10 hidden neurons and a value of $M = 12$ which results in $4 \times 12 + 1 = 49$ input neurons, similar to Bohte et al. (2002a) and Xin and Embrechts (2001). The other configuration employs 8 hidden neurons and a value of $M = 4$ which results in $4 \times 4 + 1 = 17$ input neurons (Fig. 14.6), similar to McKennoch et al. (2006). In this research, the latter configuration yielded slightly higher classification accuracies by 1-2%. As such, the accuracies are considered comparable. The second configuration, however, is preferred because it uses fewer synaptic weights, thus leading to a reduced computational requirement.

A configuration employing one exclusively inhibitory neuron in the hidden layer was investigated with SpikeProp only. As observed in the case of the XOR problem, SpikeProp converges within 500 epochs when a very low learning rate of $\eta = 0.001$ is used but not when a value of $\eta = 0.01$ is used. This configuration was not investigated further with QuickProp and RProp in this research.

14.3.4 Convergence Criteria: MSE and Training Accuracy

Due to the rounding involved in the output encoding, MSE is not an accurate measure of SNN training accuracy. Assuming that the SNN output for any training instance is encoded as one of the three prescribed output spike times of 15, 20, and 25 ms, an output spike has to be off by at least 3 ms in order to be classified incorrectly. For instance, an output of 20 ms after the rounding needs to be in the range 18-22 ms before the rounding. If it is off by 3 ms before the rounding, i.e., 17 ms or 23 ms, then it would be classified incorrectly as 15

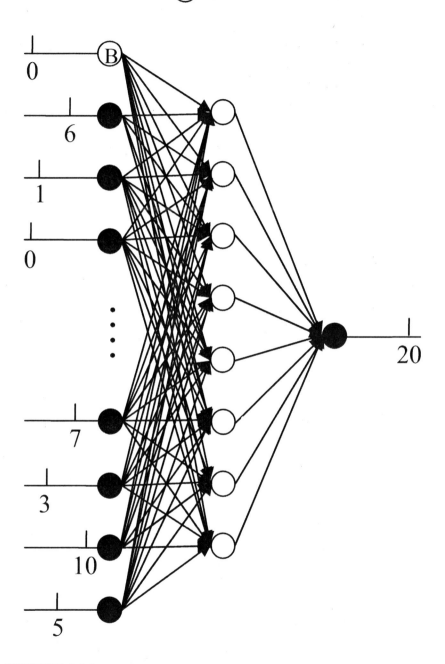

FIGURE 14.6
SNN architecture and an input-output encoding example for the Fisher iris classification problem

or 25 ms, respectively. Consider the following two cases: 1) output spikes for 4 out of, say, 100 training instances are off by 3 ms and 2) output spikes for 36 out of 100 training instances are off by 1 ms. Using Eq. (13.4) to compute the total error over all training instances and dividing the result by the number of training instances (=100), an MSE of $(1/2) \times 4 \times 3^2/100 = 0.18$ is obtained in the first case and $(1/2) \times 36 \times 1^2/100 = 0.18$ in the second case. Although the MSE values are identical, 4 spikes are classified incorrectly in the former case (96% training accuracy) whereas none are classified incorrectly in the latter case (100% training accuracy). Clearly, for an MSE of 0.18, the first case is the worst case scenario in terms of the training accuracy. In other words, selecting MSE = 0.18 implies that the network will yield a training accuracy of at least 96%.

In this work, the minimum desired training accuracy is selected as 95%. The corresponding limiting convergence MSE is computed as $(1/2) \times (100 - 95) \times 3^2/100 = 0.225$. This ensures a training accuracy of at least 95%. A rounded limiting convergence MSE of 0.25 (half of the 0.5 value selected for the XOR case) is used.

14.3.5   Heuristic Rules for Adaptive Simulation Time and SpikeProp Learning Rate

The simulation time was initially selected as 30 ms, somewhat greater than the last desired output firing time of 25 ms. However, the batch processing algorithms, QuickProp and RProp, failed to converge in some cases due to the silent neuron problem. Increasing the simulation time to 35 ms solves the problem but decreases the computational efficiency significantly due to the large size of the dataset. Therefore, the strategy used in this research is to keep the simulation time fixed at 30 ms and temporarily increase it to 35 ms whenever the network encounters the silent neuron problem. During this process of the network adjusting to silent neurons, the error values often become very large. It is observed sometimes in such a situation that SpikeProp takes a long time to converge. An adaptive error-dependent learning rate equal to 1% of MSE with an upper limit of 0.05 is employed for faster convergence. This heuristic also reduces oscillations at low MSE values where learning rates less than 0.01 reportedly perform better (Bohte et al., 2002a).

14.3.6 Classification Accuracy and Computational Efficiency versus Training Size

Table 14.4 summarizes the results of classification accuracy and computational efficiency obtained with the three learning algorithms. Each simulation is repeated 10 times with 10 different initialization weight seeds and 10 randomly selected training datasets. The training size is varied from 30 (10 training samples in each class) to 120 (40 training samples in each class) in increments of 30 (10 training samples in each class) to investigate the learning capabilities of the learning algorithms. The same learning rates that yielded the best performance for the XOR problem are used for the iris classification problem. The only exception is the value of η for QuickProp which was reduced to 0.001 because the algorithm did not converge for training size, N_T, greater than 30 for larger values. It is observed that even with this reduced learning rate QuickProp did not converge within 500 epochs for training sizes greater than 90 (data not shown in Table 14.4). Figure 14.7 shows the classification accuracies of the three learning algorithms versus the training size. Figure 14.7 and Table 14.4 show that RProp is consistently the most efficient and accurate of the three methods.

SpikeProp with incremental processing behaves differently from the two batch processing algorithms; its classification accuracy decreases with an in-

TABLE 14.4
Classification accuracy (CA) and computational efficiency (CE) using various training sizes for the iris classification problem

Training Size	SpikeProp			QuickProp			RProp		
	CA	Epoch	CE	CA	Epoch	CE	CA	Epoch	CE
30	92.7	37	7.7×10^{-4}	85.2	103	2.8×10^{-4}	90.3	31	9.2×10^{-4}
45	91.5	132	2.2×10^{-4}	91.4	53	5.4×10^{-4}	93.4	91	3.1×10^{-4}
60	91.9	111	2.6×10^{-4}	91.0	95	3.0×10^{-4}	94.8	42	6.8×10^{-4}
75	85.2	43	3.6×10^{-4}	92.3	52	5.5×10^{-4}	93.2	38	7.5×10^{-4}
90	86.2	84	3.4×10^{-4}	91.7	198	1.4×10^{-4}	93.5	40	7.1×10^{-4}
105	80.2	31	9.2×10^{-4}	-	-	-	94.7	48	6.0×10^{-4}
120	75.7	30	9.5×10^{-4}	-	-	-	94.7	49	5.8×10^{-4}

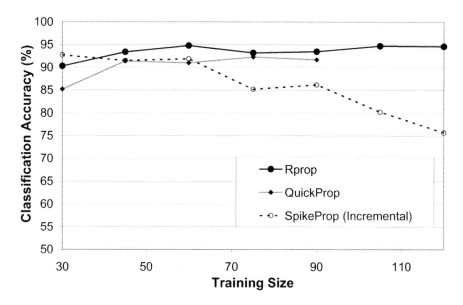

FIGURE 14.7
Classification accuracy versus training size for the Fisher iris classification
problem using three learning algorithms

crease in training size (Fig. 14.7). However, it yields comparable accuracy at
low training sizes.

14.3.7   Summary

The input and output encoding schemes used in this research result in a
reduced number of synaptic weights by a factor of 4 and simulation time by a
factor of 2, and an increase in the time step by a factor of 100 compared with
Bohte et al. (2002a). The convergence rates in Table 14.4 (for a limiting MSE
of 0.25) are similar to those reported in literature (approximately 15 epochs
for a limiting convergence MSE of 0.7). Overall, the computational effort is
decreased by a factor of $2 \times 4 \times 100 = 800$. The highest average classification
accuracy value of 94.8% is obtained using RProp. Due to the inconsistent

298 *Automated EEG-Based Diagnosis of Neurological Disorders*

performance and non-convergence of QuickProp, only SpikeProp and RProp are investigated for the EEG classification problem.

14.4 EEG Classification Problem

14.4.1 Input and Output Encoding

The mixed band EEG classification problem consists of the nine features and three classes as described in Chapter 8. None of the three classes is linearly separable. The SNN model is applied to this complex problem with a large dataset of 300 samples using SpikeProp and RProp. The same population encoding scheme used for the iris classification problem is employed for this problem to convert the continuous and real-valued inputs to discrete input spike times. An issue to be decided is the number of Gaussian functions (M) for encoding the input. A large number for M results in a proportionately large number of SNN input neurons which increases the computational requirement. In this research, four different values of M in the range 4 to 7 are investigated with the goal of maximizing the classification accuracy. A value of $\gamma = 1.5$ is used for the adjustment factor in the spread computation of the Gaussian function, the same value used for the iris classification problem.

Only one neuron is used in the output layer. The three classes are represented by the three spike times: 15, 20, and 25 ms. These values are selected by trial and error toward the middle of the range of output spike times, 2 ms to 42 ms. This range is the same as that for the iris classification problem because the input ranges and the modeling of synaptic delays are identical.

14.4.2 SNN Architecture and Training Parameters

As a result of population encoding, M input neurons are required for each of the nine input features plus a bias neuron, resulting in a total of $9M + 1$ input neurons. Corresponding to the four values of M investigated in this research, the number of SNN input neurons (including the bias neuron) used are $9 \times 4 + 1 = 37$, $9 \times 5 + 1 = 46$, $49 \times 6 + 1 = 55$, and $9 \times 7 + 1 = 64$. Two different numbers of neurons in the hidden layer were investigated: 8 and 12. Use of the larger number did not improve the classification accuracies. Consequently, the results presented in the chapter are for 8 hidden neurons.

Similar to the SNN model used for the iris classification problem, a simulation time of 30 ms is selected. The simulation time is increased temporarily to 35 ms when the silent neuron problem is encountered. The same time step of 1 ms, time decay constant of $\tau = 11$ ms, and SpikeProp learning rate of $\eta = 0.01$ are used as for the iris classification problem. It was found that the RProp learning rates used previously ($\eta^+ = 1.3$ and $\eta^- = 0.8$) were too large for this problem and the network did not converge for large training sizes. Therefore, the values of the learning rates were reduced to $\eta^+ = 1.0$ and $\eta^- = 0.7$. The results presented in the chapter are based on these values.

14.4.3 Convergence Criteria: MSE and Training Accuracy

The minimum desired training accuracy is selected as 95%. Since the output encoding for the EEG classification problem is selected to be the same as that for the iris classification problem, the limiting convergence MSE corresponding to the desired training accuracy of 95% is 0.225, as explained previously. Therefore, a rounded limiting convergence MSE of 0.25 is used. Further, for the EEG classification problem, a higher value of MSE $= 1.0$ (desired minimum training accuracy $= 77.8\%$) is also used to study the trade-off between classification accuracy and computational efficiency.

14.4.4 Classification Accuracy versus Training Size and Number of Input Neurons

The training size is varied from $3N_T = 30$ ($N_T = 10$ EEGs in each class) to $3N_T = 150$ ($N_T = 50$ EEGs in each class) in increments of 30 (10 EEGs in each class) in order to investigate the variation in classification accuracy. The variation of the classification accuracy versus the training size using Spike-Prop is shown in Fig. 14.8 for two different numbers of input neurons, 55 (corresponding to $M = 6$) and 64 (corresponding to $M = 7$). In both cases it is observed that the classification accuracy first increases and then decreases with the training size. A larger accuracy is obtained with a larger number of input neurons. Employing 64 input neurons yields the highest classification accuracy of 89.5% with a training size of 90 EEGs. The classification accuracies using 55 input neurons for training sizes greater than 90 EEGs and 64 input neurons for training sizes greater than 120 EEGs are much lower than the values shown in Fig. 14.8 and, therefore, not included.

RProp performs much better than SpikeProp for larger training sizes with similar computational efficiencies. As the training size is increased, the classification accuracy increases and plateaus near 92% for training size greater than 90 EEGs (Fig. 14.9). Increasing the number of input neurons does not affect the RProp classification accuracy significantly (not shown in Fig. 14.9).

The decrease in classification accuracy with an increase in training size observed for SpikeProp is surprising and undesirable. The behavior (also observed for the iris classification problem) is consistent with that observed during overtraining but this is not expected in this situation. It appears that the incremental nature of weight updates in SpikeProp plays a role possibly in two ways. First, in each epoch SpikeProp updates weights N_T times (equal to the training size). Therefore, as the training size increases, the number of updates increases. In contrast, in RProp, the weights are updated once every epoch

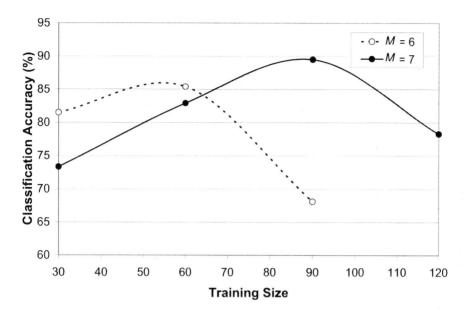

FIGURE 14.8
Classification accuracy versus training size using SpikeProp for the EEG classification problem

irrespective of training size. The increased number of weight updates repeatedly with the same training instances may lead to overtraining in SpikeProp. Second, in SpikeProp each weight update is decided by the network error from one training instance whereas in RProp the cumulative network error from all training instances is employed. The latter may lead to the network converging to a minimum that is more *generalized* and therefore yields good generalization or classification accuracy.

14.4.5 Classification Accuracy versus Desired Training Accuracy

It is noteworthy that for SpikeProp when the convergence MSE was increased from 0.25 (minimum desired training accuracy = 95.0%) to 1.0 (minimum desired training accuracy = 77.8%), the computational efficiency improved, as

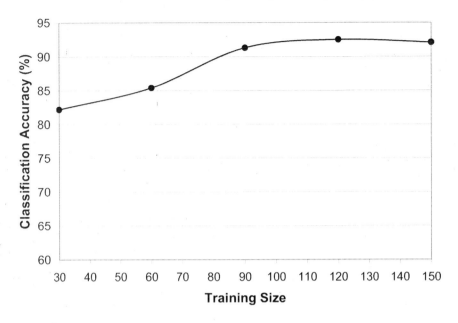

FIGURE 14.9
Classification accuracy versus training size using RProp ($M = 5$) for the EEG classification problem

expected, due to a reduction in the number of convergence epochs. However, the classification accuracy was reduced drastically to the range 30-50%. In contrast, for RProp, the classification accuracy remained the same, around 92% using MSE = 1.0. However, the computational efficiency is doubled as the number of convergence epochs is reduced by half to 28.

14.4.6   Summary

Using RProp with 46 input neurons ($M = 5$) and convergence MSE = 1.0 yields the highest classification accuracy of 92.5% and convergence rate of 28 epochs with a training size of 120 EEGs. Although RProp and SpikeProp have comparable computational efficiencies, RProp yields significantly higher classification accuracies, especially with larger training sizes. RProp is robust with respect to changes in training data and initialization weight seeds as inferred from the low standard deviations of 1.9 and 1.6, respectively.

14.5 Concluding Remarks

In this research three classification problems, XOR, Fisher iris, and EEG were investigated using three different SNN learning algorithms (SpikeProp, Quick-Prop, RProp). SpikeProp was investigated with both incremental and batch processing modes. Extensive parametric studies were performed to discover heuristics and find optimum values of the parameters with the goal of improving the computational efficiency and classification accuracy of the learning algorithms. Computational efficiency is increased by decreasing the simulation time, increasing the time step, and decreasing the number of convergence epochs. The result is a remarkable increase in computational efficiency. For the XOR problem, the computational efficiency of SpikeProp, QuickProp, and RProp is increased by a factor of 588, 82, and 75, respectively, compared with the results reported in the literature.

For the small XOR problem, it is observed that the three learning algorithms, SpikeProp (with incremental and batch processing modes), Quick-Prop, and RProp, have comparable computational efficiencies. As more complex real-world classification problems are explored, the size of the training dataset increases, which leads to a deterioration of the convergence of the batch processing algorithms except RProp. SpikeProp with incremental processing and RProp have similar computational efficiencies for the more complex iris and EEG classification problems. However, RProp yields higher classification accuracies and its classification accuracy increases with an increase in the size of the training dataset. An opposite trend is observed with Spike-Prop, an undesirable trait.

Based on extensive analyses performed in this research it is concluded that out of the three algorithms investigated RProp is the best learning algorithm because it has the highest classification accuracy, especially for large size training datasets, with about the same computational efficiency provided by SpikeProp. The error surface for the SNN model is uneven, which can be problematic for stable weight adjustments in learning algorithms. Improved classification accuracy by RProp can be explained by the fact that its weight adjustments do not depend on the magnitude of the gradients of the error surface. The SNN model for EEG classification and epilepsy and seizure detection uses RProp as the learning algorithm. This model yields a high classification accuracy of 92.5%.

15

A New Supervised Learning Algorithm for

Multiple Spiking Neural Networks

15.1 Introduction

An important characteristic of a biological presynaptic neuron is its ability
to affect a postsynaptic neuron differentially over time by inducing PSPs of
varying magnitudes at various time instants. In a biological system, this ability is incorporated using a combination of two strategies: 1) multiple spikes
at different times (*spike train*) from the presynaptic neuron to the postsynaptic neuron and/or 2) multiple synapses between two neurons with different
synaptic weights and delays. The SNN model used by Bohte et al. (2002a)
and other researchers extending their work (Xin and Embrechts, 2001; Moore,
2002; Schrauwen and van Campenhout, 2004; Silva and Ruano, 2005; McKennoch et al., 2006; Ghosh-Dastidar and Adeli, 2007) is a simplified model in
which only the latter strategy, i.e., multiple synapses, was employed and each
presynaptic neuron was restricted to the emission of a single output spike.
This model will be referred to henceforth as the single-spiking SNN model.

Due to the single-spike restriction for the SNN neuron output, information is primarily encoded with the *time to first spike*. The multiple synapses
(with different delays) per connection, however, enable the encoding to retain

a temporal aspect. In reality, these multiple synapses perform the function of modeling the spike train rather than modeling the biological aspect of the neuronal connections. In the authors' opinion, a more realistic implementation is multiple synapses per connection where every synapse has the ability to transmit spike trains (rather than single spikes). Ideally, the number of synapses per connection would also be an adaptive and *learnable* parameter. The problem with this strategy of using spike trains is a familiar one. It appeared that it would be impossible to extend the BP-based learning algorithms to such a model. Kaiser and Feldbusch (2007) attempted to circumvent this problem entirely by means of a compromise between rate encoding and pulse encoding by using a rate encoding activation function over discrete time periods. Their argument was that individual spikes can be modeled using an infinitesimally small time period that would contain only a single spike.

New learning algorithms have recently been developed that directly adapt the gradient descent based training algorithm for SNNs that convey information in the form of spike trains instead of single spikes (Booij and Nguyen, 2005; Ghosh-Dastidar and Adeli, 2009a). In these models multiple synapses per connection are no longer necessary because the spike trains inherently introduce the temporal component of PSP induction. Multiple synapses are used nevertheless because, together with spike train communication, they represent a general case similar to biological neurons. As mentioned earlier, if the number of synapses is a learnable parameter, each pair of neurons in the network can have a different number of synapses. The biological realism of SNN learning would be further advanced as synapses between neurons are added or removed as required by the learning process. The adaptive adjustment of the number of synapses similar to the model proposed by Schrauwen and van Campenhout (2004) holds significant potential and should be investigated further.

Booij and Nguyen (2005) argued that the SpikeProp algorithm and its variants were, in principle, applicable to recurrent architectures (in addition to simple feedforward architectures) for SNNs. The only restriction was that the input neurons should not be postsynaptic to any neurons and the output neurons should not be presynaptic to any neurons. They investigated the ability of their learning algorithm to learn two benchmark classification problems. One was the classical XOR problem. The other was the classification of Poisson spike trains. Two spike trains were generated using Poisson processes (Booij and Nguyen, 2005). Varying degrees of random noise were added to the two spike trains to create ten noisy versions of each spike train. The twenty spike trains were divided into training and testing sets and input to the SNN. Their learning algorithm achieved 89% accuracy in identifying the original noise-free spike trains when presented with the noisy version as input.

In this chapter, a new supervised learning or training algorithm is presented for multiple spiking neural networks that is based on more plausible neuronal dynamics (i.e., pulse-encoding spike-train communication) compared to traditional rate-encoding based ANNs and single-spiking SNNs. The new SNN model is called *multi-spiking neural network* (*MuSpiNN*) in which the presynaptic neuron transmits information to the postsynaptic neuron in the form of multiple spikes via multiple synapses. The new learning algorithm for training MuSpiNN is dubbed *Multi-SpikeProp* (Ghosh-Dastidar and Adeli, 2009a).

The output spike train of the ith presynaptic neuron input to the jth postsynaptic neuron, the resultant PSPs, and the output spike train of the jth postsynaptic neuron are shown in Fig. 15.1. In this figure, the superscript (g) indicates the sequence of the particular spike, and the time of the gth output spike of the presynaptic neuron i is denoted by $t_i^{(g)}$. For example, the time of the first output spike of the presynaptic neuron i is denoted by

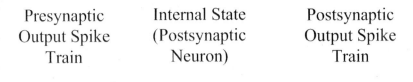

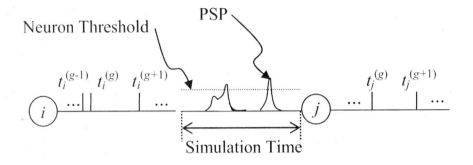

FIGURE 15.1

The output spike train of the *i*th presynaptic neuron input to the *j*th postsynaptic neuron, the resultant PSPs, and the output spike train of the *j*th postsynaptic neuron

$t_i(g = 1)$ or $t_i^{(1)}$. The same notation is used for the postsynaptic neuron j except that the subscript is changed to j.

The SNN model and learning algorithm discussed in this chapter differ from those of Booij and Nguyen (2005) in three key aspects: 1) the dynamics of the spiking neuron, 2) the learning algorithm, and 3) the heuristic rules and optimum parameter values discovered by the authors that improve the computational efficiency of the underlying SpikeProp algorithm by two orders of magnitude, as summarized in Table 15.1 (Ghosh-Dastidar and Adeli, 2007). The performance of MuSpiNN and Multi-SpikeProp is evaluated using three different problems: the XOR problem, the Fisher iris classification problem, and the EEG classification problem for epilepsy and seizure detection. For the iris and EEG classification problems, a modular architecture is employed to reduce each 3-class classification problem to three 2-class classification problems with the goal of improving the classification accuracy.

TABLE 15.1

Computational efficiency of the SpikeProp learning algorithm for the single-spiking SNN reported in the literature for the XOR problem using a simulation time of 50 ms. [*computational efficiency = time step / (simulation time × no. of epochs for convergence)*]

	Time Step	Limiting Convergence MSE	No. of Epochs for Convergence	Computational Efficiency
Bohte et al. (2002a)	0.01	0.25	250	0.8×10^{-6}
Moore (2002)	0.01	0.5	121	1.7×10^{-6}
McKennoch et al. (2006)	0.025	0.5	127	3.9×10^{-6}
Ghosh-Dastidar and Adeli (2007)	0.025	0.5	39	1.0×10^{-3}

15.2   Multi-Spiking Neural Network (MuSpiNN) and Neuron Model

15.2.1   MuSpiNN Architecture

The architecture of MuSpiNN is shown in Fig. 15.2(a). In contrast to traditional feedforward ANNs where two neurons are connected by one synapse only, the connection between two MuSpiNN neurons is modeled by multiple synapses, as shown in Fig. 15.2(b). In this aspect, the MuSpiNN model is similar to the single-spiking SNN described by Natschläger and Ruf (1998) and Bohte et al. (2002a), as discussed in Chapter 13 (Fig. 13.7). The fundamental difference is in the ability of a MuSpiNN neuron to assimilate multiple input spikes from presynaptic neurons and emit multiple output spikes in response. In other words, information transmitted from one neuron to the next is encoded in the form of a spike train instead of a single spike. The magnified connection in Fig. 15.2(b) displays the temporal sequence of spikes (short vertical lines) from the presynaptic neuron, the synaptic weights (proportionate to the size of the star shaped units in the center), and the resulting PSPs (proportionate to the size of the PSP).

The network is assumed to be fully connected, i.e., a neuron in any layer l is connected to all neurons in the preceding layer $l + 1$ (layers are numbered backward starting with the output layer, numbered as layer 1). Consequently, a neuron $j\,(\in \{1, 2, .., N_l\})$ in layer l is postsynaptic to $N_l + 1$ presynaptic neurons, where N_l is the number of neurons in layer l. Each presynaptic neuron $i\,(\in \{1, 2, .., N_{l+1}\})$ is connected to the postsynaptic neuron j via K synapses. The number K is constant for any two neurons. The weight of the kth synapse $k\,(\in \{1, 2, .., K\})$ between neurons i and j is denoted by w_{ij}^k. Assuming that presynaptic neuron i fires a total of G_i spikes and the gth spike ($g \in [1, G_i]$) is

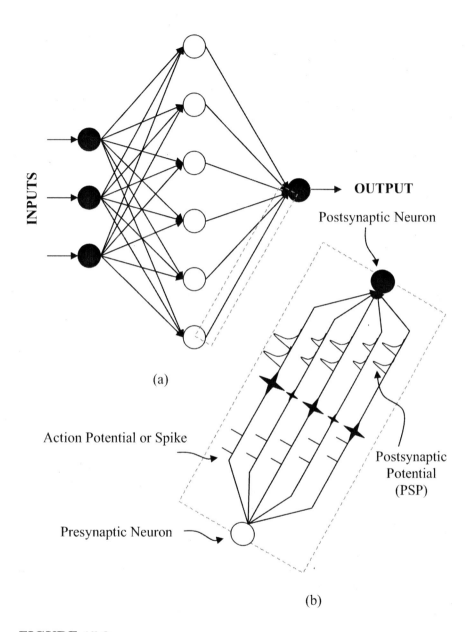

(a)

(b)

FIGURE 15.2

(a) Spiking neural network architecture; (b) multiple synapses transmitting multiple spikes from a presynaptic neuron to a postsynaptic neuron fired at time $t_i^{(g)}$ (Fig. 15.1), the kth synapse transmits the gth spike to the postsynaptic neuron at time $t_i^{(g)} + d^k$ where d^k is the delay associated with the kth synapse. The modeling of synapses is identical for all neurons, and the kth synapse between any two neurons has the same delay, d_k. Following the same notation, the output of the postsynaptic neuron j is a sequence of G_j spikes, in which the gth spike ($g \in [1, Gj]$) is fired at time $t_j^{(g)}$ (Fig. 15.1).

15.2.2   Multi-Spiking Neuron and the Spike Response Model

As long as the internal state of a biological postsynaptic neuron does not exceed the neuron threshold (Fig. 15.3), the internal state is defined as the sum of the PSPs induced by all input spikes from all presynaptic neurons and synapses. Mathematically, the internal state of the postsynaptic neuron j in layer l at time t is modeled as (Gerstner and Kistler, 2002):

$$x_j(t) = \sum_{i=1}^{N_{l+1}} \sum_{k=1}^{K} \sum_{g=1}^{G_i} w_{ij}^k \epsilon(t - t_i^{(g)} - d^k) \qquad (15.1)$$

where ϵ represents the spike response function, i.e., the PSP or the unweighted internal response of the postsynaptic neuron to a single spike. The three summations represent the weighted sum over all (G_i) input spikes from all (N_{l+1}) presynaptic neurons to the jth neuron in layer l via all (K) synapses. Zero internal state is called the resting potential (Fig. 15.3). Similar to the spike response model for the SNN discussed in Chapter 13, the α-function described by Eq. (13.2) is used as the spike response function (Bohte et al., 2002a; Gerstner and Kistler, 2002). Booij and Nguyen (2005) use a different spike response function defined as the difference of two exponential functions. The model and the learning algorithm presented in this chapter can also incorporate that function but no significant changes in the result are expected.

When the internal state exceeds the neuron threshold, θ, the neuron fires

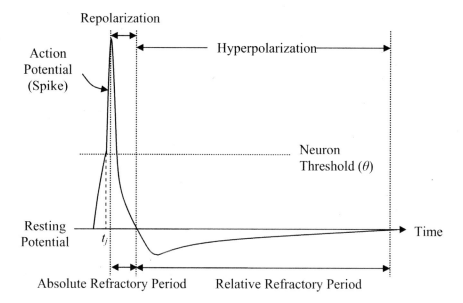

FIGURE 15.3
The internal state of a postsynaptic neuron in response to a presynaptic spike (not shown in the figure) showing the action potential, and repolarization and hyperpolarization phases.

an output spike at the time instant t_j, and the internal state immediately starts dropping to the resting potential of the neuron (Fig. 15.3). This is the *repolarization* phase (Fig. 15.3). The duration of the repolarization phase is known as the *absolute refractory period* in which the neuron cannot fire regardless of the number or frequency of the input spikes (Fig. 15.3). Subsequently, the internal state is kept at a value lower than the resting potential by various biological processes. As a result, it becomes difficult for the neuron to reach the threshold and fire again for a period of time, known as the *relative refractory period*. This is the *hyperpolarization* phase (Fig. 15.3) (Bose and Liang, 1996; Kandel et al., 2000). Due to repolarization and hyperpolarization, the internal state of a postsynaptic neuron depends not only on the timing of the input spikes from all presynaptic neurons but also on the timing of its own output spikes. A more detailed discussion of the modeling of the internal state of a postsynaptic neuron is presented in Chapter 13.

Since a single-spiking neuron is restricted to firing only one output spike, the single-spiking SNN model of Bohte et al. (2002a) is unaffected by the timing of its own output spikes. Eqs. (15.1) and (13.2) are sufficient to represent the dynamics of a single-spiking neuron but not for the multi-spiking neuron in the new MuSpiNN model. To model the relative refractory period for the multi-spiking neuron, a refractoriness term is added to the right-hand side of Eq. (15.1). This refractoriness term ensures that the membrane potential becomes negative after the firing of a spike, which makes it difficult for the neuron to emit subsequent spikes for a period of time, as explained earlier. Therefore, in this chapter the internal state of the postsynaptic neuron j in layer l at time t is expressed as (Gerstner and Kistler, 2002):

$$x_j(t) = \sum_{i=1}^{N_{l+1}} \sum_{k=1}^{K} \sum_{g=1}^{G_i} w_{ij}^k \epsilon(t - t_i^{(g)} - d^k) + \rho(t - t_j^{(f)}) \qquad (15.2)$$

where ρ represents the refractoriness function and $t_j^{(f)}$ is the timing of the most recent, the fth, output spike from neuron j prior to time t. For $t < t_j^{(1)}$, the time of the first output spike, the refractoriness term is zero and Eq. (15.2) is reduced to Eq. (15.1). Equation (15.2) is different from the corresponding equation presented in the model of Booij and Nguyen (2005) where the refractoriness term is summed over all output spikes from neuron j prior to time t (instead of only the most recent one). That model assumes that at any time t, the internal state of the neuron is affected by the refractoriness due to all spikes prior to time t. This assumption is biologically unrealistic because for a biological neuron, the ensuing refractoriness after every spike has to be overcome before the neuron can spike again. Therefore, when the neuron does spike the next time, it is implicit that the preceding refractoriness has been overcome and is of no further consequence. From this progression of events it is clear that, at any time, no refractoriness is retained from any previous spike except the most recent one, which is reflected in Eq. (15.2).

In this chapter, ρ is expressed as:

$$\rho(t) = \begin{cases} -2\theta e^{t/\tau_R} & \text{when} \quad t > 0 \\ 0 & \text{when} \quad t \leq 0 \end{cases} \tag{15.3}$$

where τ_R is the time decay constant that determines the spread shape of the refractoriness function. Figure 15.4 shows the refractoriness function for three different values of τ_R. The function has a negative value in the range $t = 0$ to ∞ with a minimum value of -2θ at $t = 0$. Its value increases with time and approaches zero at $t = \infty$. Substituting Eq. (15.3) in Eq. (15.2) ensures that at the instant of spike firing, $t_j^{(f)}$, the internal state of the neuron decreases instantaneously from the threshold θ to $\theta - 2\theta = -\theta$. Subsequently, the internal state remains lower than the resting potential and increases exponentially to the resting potential which accurately models the relative refractory period and the hyperpolarization phase shown in Fig. 15.3. The corresponding equation in Booij and Nguyen (2005) uses a coefficient of $-\theta$ for the exponential term in Eq. (15.3) which reduces the internal state to the resting potential (but not lower) at the instant of spike firing. At no subsequent point in time does the internal state decrease to a value less than the resting potential, which does not accurately model the relative refractory period and the hyperpolarization phase (Fig. 15.3). Figure 15.5 shows the mathematical model of the overall neuron dynamics represented by Eq. (15.1). This model is an approximate representation of the dynamics of the biological neuron shown in Fig. 15.3. The absolute refractory period observed in the biological model (Fig. 15.3) is neglected in the mathematical model.

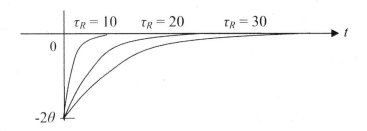

FIGURE 15.4

The refractoriness function ρ for three different values $\tau_R = 10$, 20, and 30.

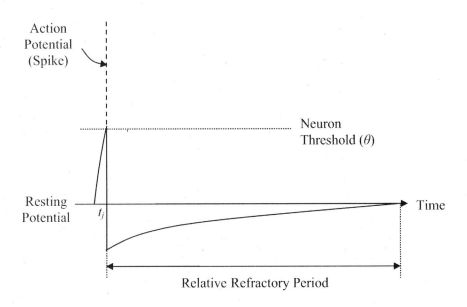

FIGURE 15.5

The mathematical model of the internal state of a postsynaptic neuron in response to a presynaptic spike (not shown in the figure) showing the action potential and the relative refractory period.

15.3   Multi-SpikeProp: Backpropagation Learning Algorithm for MuSpiNN

15.3.1   MuSpiNN Error Function

Supervised learning requires the availability of a set of actual or *desired* output spike trains which is the desired output from MuSpiNN given a set of input spike trains (one spike train per input neuron). A measure of the difference between the computed and desired outputs is used to compute the network error. MuSpiNN is trained by backpropagating the error (as explained shortly) and adjusting the synaptic weights such that the network error is minimized. Desired outputs for SNNs, unlike those for traditional neural networks, have to be in terms of discrete spike times. The transformation of real-valued outputs to discrete spike times is known as *output encoding* and it can be achieved in a number of different ways, as described in Chapter 14. The appropriateness of any particular output encoding depends on the specific problem application.

In this chapter, neurons in the output layer of MuSpiNN are restricted to emitting a single output spike so that the error function is computed with no additional difficulty. The availability of multiple output spikes at various discrete times requires a more complicated error function. The complexity is compounded by the fact that the number of spikes in the computed output spike trains is variable and highly likely to be different from the number in the desired output spike trains. Such an error function needs to be explored in the context of the output encoding for selected problems, which is beyond the scope of this book. As such, considering only one output spike, the network error is computed as follows:

$$E = \frac{1}{2}\sum_{j=1}^{N_1}(t_j - t_j^d)^2 \qquad (15.4)$$

where t_j and t_j^d are the computed and desired spike times, respectively, for the jth neuron in the output layer $(l = 1)$.

15.3.2   Error Backpropagation for Adjusting Synaptic Weights

The generalized delta update rule is employed to backpropagate the error and adjust the synaptic weights. The weight adjustment for the kth synapse between the ith presynaptic and the jth postsynaptic neuron is computed as:

$$\Delta w_{ij}^k = -\eta \nabla E_{ij}^k \qquad (15.5)$$

where η is the learning rate and ∇E_{ij}^k is the gradient (with respect to the weights) of the error function for the kth synapse between the ith presynaptic and jth postsynaptic neuron. The computation of the gradient is different for the output layer and the hidden layers and is described separately for the two in the following sections. For the sake of clarity, a neuron in the output layer $(l = 1)$ will be designated with the subscript j and a neuron in the hidden layer immediately presynaptic to the output layer $(l = 2)$ with the subscript i. A neuron in the input or hidden layer presynaptic to the hidden layer $l = 2$ is designated by the subscript h. The subscripts of all other variables are adjusted based on this nomenclature.

15.3.3 Gradient Computation for Synapses Between a Neuron in the Last Hidden Layer and a Neuron in the Output Layer

Using the chain rule the error gradient at the postsynaptic neuron output spike time instant, $t = t_j$, is represented as the product of three partial derivative terms:

$$\nabla E_{ij}^k = \frac{\partial E}{\partial w_{ij}^k}$$
$$= \frac{\partial E}{\partial t_j} \frac{\partial t_j}{\partial x_j(t_j)} \frac{\partial x_j(t_j)}{\partial w_{ij}^k} \tag{15.6}$$

The first, third, and second partial derivative terms are derived in that order, which is the order of their complexity. The first partial derivative term on the right-hand side is computed as:

$$\frac{\partial E}{\partial t_j} = \frac{\partial \left[\frac{1}{2} \sum\limits_{j=1}^{N_1} (t_j - t_j^d)^2 \right]}{\partial t_j}$$
$$= (t_j - t_j^d) \tag{15.7}$$

The third partial derivative term on the right-hand side of Eq. (15.6) is computed as:

$$\frac{\partial x_j(t_j)}{\partial w_{ij}^k} = \frac{\partial \left[\sum\limits_{i=1}^{N_2} \sum\limits_{k=1}^{K} \sum\limits_{g=1}^{G_i} w_{ij}^k \epsilon(t_j - t_i^{(g)} - d^k) + \rho(t_j - t_j^{(f)}) \right]}{\partial w_{ij}^k} \tag{15.8}$$

Since the neurons in the output layer are permitted to fire only one output spike, the refractoriness term in the numerator of Eq. (15.8) becomes zero. Because the weight of any one synapse is independent of the weights of the other synapses, the summations with respect to i and k vanish and Eq. (15.8) is reduced to:

$$\frac{\partial x_j(t_j)}{\partial w_{ij}^k} = \frac{\partial \left[\sum\limits_{g=1}^{G_i} w_{ij}^k \epsilon(t_j - t_i{}^{(g)} - d^k) \right]}{\partial w_{ij}^k} \tag{15.9}$$

Since the α-function, ϵ, is independent of the synaptic weights, Eq. (15.9) is further simplified to:

$$\frac{\partial x_j(t_j)}{\partial w_{ij}^k} = \sum_{g=1}^{G_i} \epsilon(t_j - t_i{}^{(g)} - d^k) \tag{15.10}$$

Eq. (15.10) is a more generalized form of the one used by Bohte et al. (2002a), which is limited to single input spikes from neurons presynaptic to the output layer neurons.

The second partial derivative term in Eq. (15.6), $\partial t_j / \partial x_j(t_j)$, cannot be computed directly because t_j cannot be expressed as a continuous and differentiable function of $x_j(t_j)$, as explained earlier. Bohte et al. (2002a) overcome this problem by assuming that $x_j(t_j)$ is a linear function of t_j around the output spike time instant, $t = t_j$, and approximate the term $\partial t_j / \partial x_j(t_j)$ as $-1/[\partial x_j(t_j)/\partial t_j]$. In this chapter, the authors arrive at the same solution but offer a different or, perhaps, a clearer explanation of this approximation.

Consider an example graph of the internal state $x_j(t)$ versus time t shown in Fig. 15.6(a). The internal state reaches threshold θ and fires an output spike at time t_j. Before reaching the threshold θ, the internal state $x_j(t)$ is independent of the threshold. A decrease in θ leads to a decrease in the output spike time t_j (the neuron fires earlier) and vice versa. This is evident from Fig. 15.6(a) where output spike time t_j occurs at the intersection of the dashed line representing θ and the solid line representing the graph of $x_j(t)$. As a result, if θ is considered as a variable parameter, any point $(t, x_j(t))$ can be represented by the point (t_j, θ). Figure 15.6(b) shows the variation of t_j with θ. It is observed that if θ is reduced to θ_1, t_j is proportionately reduced to t_{j1}.

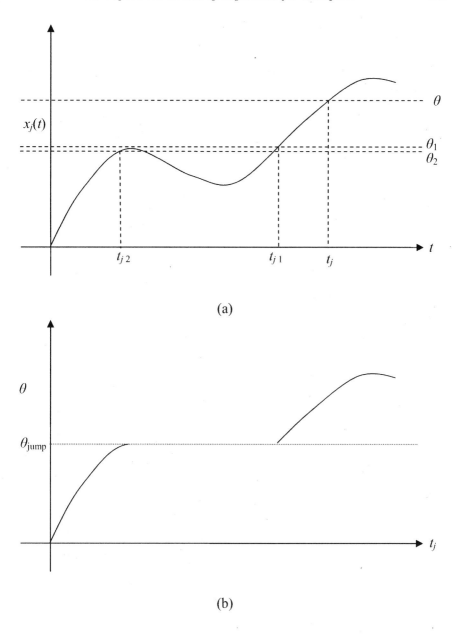

(a)

(b)

FIGURE 15.6
(a) Internal state, $x_j(t)$, versus time, t; (b) neuron threshold, θ, versus output spike time, t_j

However, when θ_1 is further reduced to θ_2, t_{j1} is disproportionately reduced to t_{j2}. The threshold θ at which the spike time jumps is designated θ_{jump} in this chapter.

Following the literature in SNN research, it is assumed that the value of the threshold θ is not close to θ_{jump}. Based on extensive modeling and simulations it has been found that limiting learning rates and weight changes during learning contains the error function locally within the error surface and avoids drastic jumps in spike times. This can be ensured by choosing a low learning rate and developing heuristic rules to be discussed later. As a result, the assumption of linearity holds around the spike time t_j where the slopes of the two graphs shown in Figs. 15.6(a) and (b) are identical, i.e.,

$$\left.\frac{\partial x_j(t)}{\partial t}\right|_{t=t_j} = \frac{\partial \theta}{\partial t_j} \qquad (15.11)$$

The left-hand side of Eq. (15.11) is approximated instantaneously as $\partial x_j(t_j)/\partial t_j$ for small changes in $x_j(t)$. As a result of the assumption of linearity around time t_j, the slope of the graph of θ versus t_j is computed as:

$$\begin{aligned}\frac{\partial t_j}{\partial \theta} &= \frac{\Delta t_j}{\Delta \theta} \\ &= \frac{1}{\Delta\theta/\Delta t_j} \\ &= \frac{1}{\partial\theta/\partial t_j} \qquad (15.12)\end{aligned}$$

where Δt_j represents an infinitesimal change in t_j and $\Delta\theta$ represents an infinitesimal change in θ. Moreover, to increase the output spike time t_j, the internal state of the neuron $x_j(t_j)$ has to be decreased or, equivalently, the threshold θ has to be increased. This opposite relationship between the internal state and threshold, with respect to the output spike time, is modeled as:

$$\frac{\partial t_j}{\partial x_j(t_j)} = -\frac{\partial t_j}{\partial \theta} \tag{15.13}$$

From Eqs. (15.11), (15.12), and (15.13), the second partial derivative term in Eq. (15.6) becomes

$$\frac{\partial t_j}{\partial x_j(t_j)} = \frac{-1}{\partial x_j(t_j)/\partial t_j} \tag{15.14}$$

The denominator of Eq. (15.14) is computed as:

$$\frac{\partial x_j(t_j)}{\partial t_j} = \frac{\partial \left[\sum_{i=1}^{N_2} \sum_{k=1}^{K} \sum_{g=1}^{G_i} w_{ij}^k \epsilon(t_j - t_i^{(g)} - d^k) + \rho(t_j - t_j^{(f)}) \right]}{\partial t_j} \tag{15.15}$$

Similar to the derivation of Eq. (15.9), since the neurons in the output layer are permitted to fire only one output spike, the refractoriness term in the numerator of Eq. (15.15) becomes zero. Because the output spike times transmitted through one synapse are independent of the output spike times through the other synapses, the summations can be placed outside the derivative and Eq. (15.15) is rewritten as:

$$\frac{\partial x_j(t_j)}{\partial t_j} = \sum_{i=1}^{N_2} \sum_{k=1}^{K} \sum_{g=1}^{G_i} w_{ij}^k \frac{\epsilon(t_j - t_i^{(g)} - d^k)}{\partial t_j}$$

$$= \sum_{i=1}^{N_2} \sum_{k=1}^{K} \sum_{g=1}^{G_i} w_{ij}^k \epsilon(t_j - t_i^{(g)} - d^k) \left(\frac{1}{(t_j - t_i^{(g)} - d^k)} - \frac{1}{\tau} \right) \tag{15.16}$$

Substituting Eq. (15.16) in Eq. (15.14), we obtain

$$\frac{\partial t_j}{\partial x_j(t_j)} = \frac{-1}{\sum_{i=1}^{N_2} \sum_{k=1}^{K} \sum_{g=1}^{G_i} w_{ij}^k \epsilon(t_j - t_i^{(g)} - d^k) \left(\frac{1}{(t_j - t_i^{(g)} - d^k)} - \frac{1}{\tau} \right)} \tag{15.17}$$

In summary, the error gradient for adjusting the synaptic weights, ∇E_{ij}^k, is computed using Eqs. (15.6), (15.7), (15.10), and (15.17).

Figures 15.6(a) and (b) can be used to explain the reason for requiring low learning rates and various heuristic rules to limit the changes of the synaptic weights (Ghosh-Dastidar and Adeli, 2007). In the example shown, assume that the weights need to be increased in order to increase the internal state of the neuron (or, equivalently, decrease the threshold) and obtain an earlier spike time. As long as the weight changes are small, the graph of θ versus t_j remains continuous and the assumption of linearity is not unreasonable. However, if the changes in the weights are too large and lead to the region around θ_{jump} in Fig. 15.6(b), a small change in θ leads to a disproportionate change in the output spike time t_j which jumps to a much earlier time. In this region, the assumption of linearity does not hold, and SNN training fails.

15.3.4 Gradient Computation for Synapses Between a Neuron in the Input or Hidden Layer and a Neuron in the Hidden Layer

For a postsynaptic neuron i in a hidden layer $l = 2$, the error is backpropagated from all neurons in the output layer, and the gradient is computed using the chain rule as:

$$\nabla E_{hi}^k = \sum_{j=1}^{N_l} \frac{\partial E}{\partial w_{hi}^k}$$
$$= \sum_{j=1}^{N_l} \frac{\partial E}{\partial t_i^{(g)}} \frac{\partial t_i^{(g)}}{\partial w_{hi}^k} \tag{15.18}$$

where the subscript h denotes the hth neuron in layer $l+1$ that is presynaptic to the postsynaptic neuron i in layer l. The first partial derivative term in Eq. (15.18) models the dependence of the network error in Eq. (15.4) on the output spike times $t_i^{(g)}$ from the neuron i in the hidden layer $l = 2$ and is expanded using the chain rule as:

$$\frac{\partial E}{\partial t_i{}^{(g)}} = \frac{\partial E}{\partial t_j} \frac{\partial t_j}{\partial x_j(t_j)} \frac{\partial x_j(t_j)}{\partial t_i{}^{(g)}} \qquad (15.19)$$

The first two partial derivative terms on the right-hand side are computed according to Eqs. (15.7) and (15.17), respectively. The last partial derivative term is computed as:

$$\frac{\partial x_j(t_j)}{\partial t_i{}^{(g)}} = \frac{\partial \left[\sum\limits_{i=1}^{N_{l+1}} \sum\limits_{k=1}^{K} \sum\limits_{g=1}^{G_i} w_{ij}^k \epsilon(t_j - t_i{}^{(g)} - d^k) + \rho(t_j - t_j{}^{(f)}) \right]}{\partial t_i{}^{(g)}} \qquad (15.20)$$

The output spike times, $t_i{}^{(g)}$, from neuron i in the hidden layer $l = 2$ are the input spike times for neuron j in the output layer $l = 1$ and independent of the output spike times for neuron j. Therefore, the refractoriness term in the derivative vanishes. The summation with respect to i also vanishes because the output spikes of any neuron in the hidden layer do not depend on the output spikes of any other neuron in the same layer. Since the synaptic weights are independent of the output spike times, the factor w_{ij}^k is placed outside the derivative and Eq. (15.20) is reduced to:

$$\frac{\partial x_j(t_j)}{\partial t_i{}^{(g)}} = \sum_{k=1}^{K} \sum_{g=1}^{G_i} w_{ij}^k \frac{\epsilon(t_j - t_i{}^{(g)} - d^k)}{\partial t_i{}^{(g)}}$$

$$= -\sum_{k=1}^{K} \sum_{g=1}^{G_i} w_{ij}^k \epsilon(t_j - t_i{}^{(g)} - d^k) \left(\frac{1}{(t_j - t_i{}^{(g)} - d^k)} - \frac{1}{\tau} \right) \qquad (15.21)$$

The negative sign in Eq. (15.21) appears because the derivative of the α-function, $\epsilon(t_j - t_i{}^{(g)} - d^k)$, is calculated with respect to $t_i{}^{(g)}$ (unlike Eq. 15.16 where it is calculated with respect to t_j).

The computation of the second partial derivative term in Eq. (15.18) is more complicated than in a single-spiking model where the output spike time

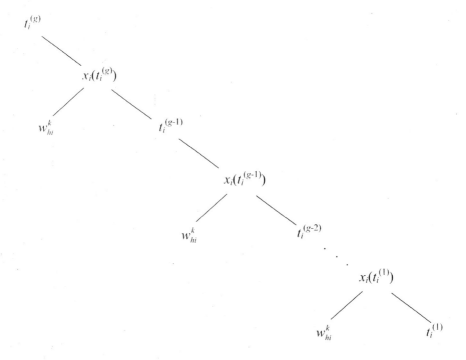

FIGURE 15.7
Recursive dependencies of the times of output spikes from the postsynaptic neuron i on the weights of synapses connecting the presynaptic neuron h to the postsynaptic neuron i

from any neuron depends on the internal state of the neuron i which, in turn, depends only on the inputs and weights of the synapses to the neuron. In the MuSpiNN model, the internal state of the neuron i, in addition to the afore-mentioned dependencies, depends on the time of its own most recent output spike. Figure 15.7 shows the recursive dependencies of the times of output spikes from the postsynaptic neuron i on the weights of synapses connecting the presynaptic neuron h to the postsynaptic neuron i where the variable represented by any node in the tree is dependent on the variables represented by the nodes in the level immediately below. As a result of these recursive dependencies, the error is backpropagated from output spike to output spike starting from the last output spike. The second partial derivative term in Eq.

(15.18) is computed recursively using the following set of equations:

$$\frac{\partial t_i^{(g)}}{\partial w_{hi}^k} = \frac{\partial t_i^{(g)}}{\partial x_i(t_i^{(g)})} \left[\frac{\partial x_i(t_i^{(g)})}{\partial w_{hi}^k} + \frac{\partial x_i(t_i^{(g)})}{\partial t_i^{(g-1)}} \cdot \frac{\partial t_i^{(g-1)}}{\partial w_{hi}^k} \right] \tag{15.22}$$

$$\frac{\partial t_i^{(g-1)}}{\partial w_{hi}^k} = \frac{\partial t_i^{(g-1)}}{\partial x_i(t_i^{(g-1)})} \left[\frac{\partial x_i(t_i^{(g-1)})}{\partial w_{hi}^k} + \frac{\partial x_i(t_i^{(g-1)})}{\partial t_i^{(g-2)}} \cdot \frac{\partial t_i^{(g-2)}}{\partial w_{hi}^k} \right] \tag{15.23}$$

$$\vdots$$

$$\frac{\partial t_i^{(1)}}{\partial w_{hi}^k} = \frac{\partial t_i^{(1)}}{\partial x_i(t_i^{(1)})} \left[\frac{\partial x_i(t_i^{(1)})}{\partial w_{hi}^k} \right] \tag{15.24}$$

Similar to the derivation of Eq. (15.17) from Eq. (15.14), the first partial derivative term outside the brackets in Eq. (15.22), $\partial t_i^{(g)}/\partial x_i(t_i^{(g)})$, is computed as:

$$\frac{\partial t_i^{(g)}}{\partial x_i(t_i^{(g)})} = \frac{-1}{\partial x_i(t_i^{(g)})/\partial t_i^{(g)}}$$

$$= -\left[\partial x_i(t_i^{(g)})/\partial t_i^{(g)} \right]^{-1}$$

$$= -\left[\partial \left(\sum_{h=1}^{N_l} \sum_{k=1}^{K} \sum_{f=1}^{G_h} w_{hi}^k \epsilon(t_i^{(g)} - t_h^{(f)} - d^k) \right. \right.$$

$$\left. \left. + \rho(t_i^{(g)} - t_i^{(g-1)}) \right) \middle/ \partial t_i^{(g)} \right]^{-1}$$

$$= -\left[\sum_{h=1}^{N_l} \sum_{k=1}^{K} \sum_{f=1}^{G_h} w_{hi}^k \frac{\partial \epsilon(t_i^{(g)} - t_h^{(f)} - d^k)}{\partial t_i^{(g)}} \right.$$

$$\left. + \frac{\partial \rho(t_i^{(g)} - t_i^{(g-1)})}{\partial t_i^{(g)}} \right]^{-1}$$

$$= -\left[\sum_{h=1}^{N_l} \sum_{k=1}^{K} \sum_{f=1}^{G_h} w_{hi}^k \epsilon(t_i^{(g)} - t_h^{(f)} - d^k) \left(\frac{1}{t_i^{(g)} - t_h^{(f)} - d^k} - \frac{1}{\tau} \right) \right.$$

$$\left. + \frac{2\theta}{\tau_R} \rho(t_i^{(g)} - t_i^{(g-1)}) \right]^{-1} \tag{15.25}$$

Similar to Eq. (15.10), the first term within the brackets in Eq. (15.22) is computed as:

$$\frac{\partial x_i(t_i^{(g)})}{\partial w_{hi}^k} = \sum_{f=1}^{G_h} \epsilon(t_i^{(g)} - t_h^{(f)} - d^k) \tag{15.26}$$

The first partial derivative term of the second term within the brackets in Eq. (15.22) is computed as:

$$\frac{\partial x_i(t_i^{(g)})}{\partial t_i^{(g-1)}} = \frac{\partial \left[\sum_{h=1}^{N_I} \sum_{k=1}^{K} \sum_{f=1}^{G_h} w_{hi}^k \epsilon(t_i^{(g)} - t_h^{(f)} - d^k) + \rho(t_i^{(g)} - t_i^{(g-1)}) \right]}{\partial t_i^{(g-1)}} \tag{15.27}$$

The first term in the numerator does not depend on any previous output spike times; thus Eq. (15.27) is reduced to:

$$\frac{\partial x_i(t_i^{(g)})}{\partial t_i^{(g-1)}} = \frac{\partial \rho(t_i^{(g)} - t_i^{(g-1)})}{\partial t_i^{(g-1)}}$$

$$= \frac{2\theta}{\tau_R} \rho(t_i^{(g)} - t_i^{(g-1)}) \tag{15.28}$$

The error gradient for adjusting the synaptic weights is computed using Eqs. (15.7), (15.17)-(15.19), (15.21)-(15.26), and (15.28).

16

Applications of Multiple Spiking Neural
Networks: EEG Classification and Epilepsy
and Seizure Detection

16.1   Parameter Selection and Weight Initialization

The performance of the MuSpiNN model and the Multi-SpikeProp learning algorithm is evaluated using three increasingly difficult pattern recognition problems: XOR, Fisher iris plant classification (Fisher, 1936; Newman et al., 1998), and EEG epilepsy and seizure detection (Andrzejak et al., 2001; Adeli et al., 2007; Ghosh-Dastidar et al., 2007, 2008; Ghosh-Dastidar and Adeli, 2007). The classification accuracy is computed only for the iris and EEG datasets because they are large enough to be divided into training and testing datasets. The transformation of real-valued inputs and outputs to discrete spike times (output encoding) is different for the three problems and therefore is addressed for each problem separately.

The performance of SNNs is affected by three types of parameters that define a) the spiking neuron (simulation time, time step, neuron threshold, time decay constant τ for the α-function, and time decay constant τ_R for the refractoriness function), b) network architecture (number of hidden layers, input and output encoding parameters, number of neurons in the input, hidden, and output layers, and number of synapses connecting two neurons), and c) the learning algorithm (learning rate and the convergence criteria). In Chapter 14, an extensive parametric analysis was presented for selecting optimum values of the aforementioned parameters for a single-spiking SNN model and SpikeProp (Ghosh-Dastidar and Adeli, 2007). The goal was to maximize the accuracy and efficiency of the model. Maximum computation efficiency and classification accuracies were achieved when a learning rate in the range of $\eta = 0.001 - 0.014$ and a time step of 1 ms were employed (the time unit is *virtual* and is used for modeling purposes only). The simulation time employed for the XOR problem was 25 ms and for the iris and EEG problem 35 ms.

The selection of these parameters has been described in detail in Chapter 14 (Ghosh-Dastidar and Adeli, 2007). In MuSpiNN, the underlying model is similar to the SNN and the difference lies primarily in the biologically more realistic learning algorithm capable of handling multiple spikes. Therefore, it is expected that the same or similar values of parameters will be optimum. The new MuSpiNN model is developed around the same optimum numbers obtained for learning rate, simulation time, time step, number of hidden layers, input and output encoding parameters, and number of neurons in the input, hidden, and output layers. The optimum values of the remaining parameters are obtained differently for the multi-spiking model. Parameter values are different for the three classification problems and, therefore, discussed separately for each problem in the following sections.

Weights for MuSpiNN training are initialized in a manner similar to that described for an SNN in Chapter 14. To increase the consistency of convergence of the network training, all neurons are required to fire within the simulation time, at least in the first epoch of network training.

16.2   Heuristic Rules for Multi-SpikeProp

The improved SNN model presented in Chapter 14 employed two heuristic rules that increased the computational efficiency and classification accuracy of the model. The same heuristic rules along with a third new rule are employed for MuSpiNN and Multi-SpikeProp as follows:

1. In order to prevent catastrophic changes in the synaptic weights, a lower limit of 0.1 is imposed on the denominator in Eqs. (15.17) and (15.25), as suggested by Booij and Nguyen (2005) for their equations, which are different from the ones described in Chapter 15.

2. If at any time during the training of the network, a neuron stops firing, then its contribution to the network error becomes null. During back-propagation of the error, the resulting weight change is very small, which may not be sufficient to restart the firing of the neuron even after several epochs. This issue, referred to as the *silent neuron problem*, leads to a reduction of the effective network size to a size possibly insufficient to model the classification problem, which ultimately affects convergence (McKennoch et al., 2006). In our model, the neuron is set to fire at the maximum internal state value if the threshold is not exceeded during the simulation time. Based on this heuristic, every neuron fires during the simulation time.

3. Our mathematical model of neuron dynamics, represented by Eq. (15.2), incorporates the relative refractory period but not the absolute refractory period (Figs. 15.3 and 15.5). As a result, in some situations during the training of the network, the large weighted inputs to a neuron offset the effect of the refractoriness function, resulting in an effective

refractory period equal to 0 ms. In that case, the neuron starts firing continuously, which causes all of its postsynaptic neurons to behave in a similar manner, thus overwhelming the network and adversely affecting convergence. This problem, dubbed the *noisy neuron problem*, was not encountered for the SNN model, which is restricted to only one spike. The problem is analogous to propagation of seizure in an epileptic brain where abnormal neuronal discharges spread via a similar mechanism, i.e., feedforward excitation, to various parts of the brain. In order to overcome this problem, a new heuristic rule is added to model the absolute refractory period by requiring that any neuron be unable to fire again within 2 ms following an output spike.

16.3   XOR Problem

The encoding of the XOR problem for MuSpiNN is identical to that for SNN, as described in detail in Section 14.2.1. The number of layers and neurons in the neural network architecture is also kept unchanged. The number of input and output neurons is selected as 3 (including the *bias* neuron) and 1, respectively. The reason for the use of the bias neuron is explained in Section 14.2.2. Only one hidden layer comprising 5 neurons is used to model the XOR problem.

The model is trained to a convergence mean square error (MSE) value of 0.5. The upper limit for the number of epochs is set to 500. The training is repeated for ten different sets of initialization weights. The learning rate η was selected as 0.005; the model did not converge for higher rates. This value of the learning rate resulted in the fastest learning. The time decay constant for the refractoriness function, τ_R, was selected as 80 ms by trial and error.

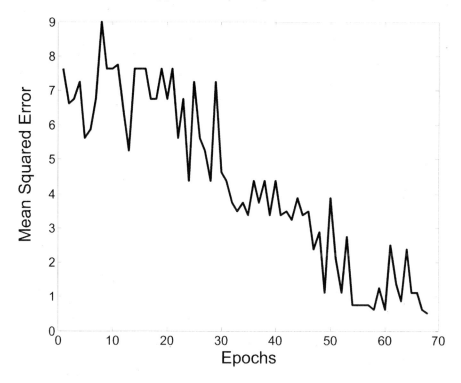

FIGURE 16.1

Sample Multi-SpikeProp convergence curve for the XOR problem

Other values of τ_R in the range 0 to 100 ms were investigated. Although the convergence was adversely affected for values lower than 20 ms, no consistent pattern was observed for values greater than that. A sample Multi-SpikeProp convergence curve for MuSpiNN is shown in Fig. 16.1.

Multi-SpikeProp converges in an average of 78 epochs, which is approximately double the 38 epochs required by SpikeProp (Ghosh-Dastidar and Adeli, 2007). However, Multi-SpikeProp requires only 4 synapses connecting a presynaptic neuron to a postsynaptic neuron, compared with 16 synapses required by SpikeProp. A number of synapses lower than 4 appears to be insufficient to model the problem and leads to less consistent convergence. A larger number of synapses, say 6, leads to faster convergence (57 epochs) but the computational effort and time are increased.

Our parametric studies indicate that the delays associated with the synapses should be spread out in the whole range between the earliest input spike (0 ms) and the latest output spike (16 ms). MuSpiNN converges in 78 epochs when delays of 1, 5, 9, and 13 ms are employed for the four synapses but fails to converge when delays of 1, 2, 3, and 4 ms or 1, 3, 5, and 7 ms are employed. The single-spiking SNN model and SpikeProp were also investigated with the same sets of delays and 4 synapses for comparison with MuSpiNN. It was observed that SpikeProp also failed to converge when delays of 1, 2, 3, and 4 ms or 1, 3, 5, and 7 ms were employed. The consistency of convergence improved when delays of 1, 5, 9, and 13 ms were employed for the four synapses but the number of convergence epochs was much greater (in excess of 200 epochs) than that required by MuSpiNN.

16.4　Fisher Iris Classification Problem

The selection of input encoding parameters and the number of neurons in each layer for MuSpiNN and Multi-SpikeProp parameters are based on the research using SpikeProp on the iris problem, as described in Section 14.3. For input encoding using population encoding, a value of $\gamma = 1.5$ yielded the best classification accuracies (Bohte et al., 2002a; Ghosh-Dastidar and Adeli, 2007). The number of input neurons (equal to the number of population encoding Gaussian functions, M) required per input feature is selected as four, resulting in a total of $4M + 1 = 4 \times 4 + 1 = 17$ neurons (including one bias neuron) in the input layer. The number of neurons in the hidden layer is selected as eight. Based on the performance of the SNN for the iris problem discussed in

Section 14.3 and MuSpiNN for the XOR problem, the time decay constant τ is selected as 11 ms (slightly larger than the encoding interval of 10 ms), the refractoriness function time decay constant τ_R as 80 ms, and the learning rate η as 0.01. Four synapses are used to connect any presynaptic neuron to a postsynaptic neuron with delays of 1, 5, 9, and 13 ms.

A modular structure composed of three MuSpiNN modules is employed for solving the three-class classification problem. Figure 16.2 shows the MuSpiNN architectures for (a) the original three-class classification problem and (b) three two-class classification problems. In Fig. 16.2(b), each MuSpiNN module is dedicated to one class and assigned the task of classifying the data as either belonging to that class or not belonging to that class. If the data belong to that class, the MuSpiNN module responds with an output spike at 15 ms, and otherwise at 20 ms. Therefore, the three-class classification problem in Fig. 16.2(a) is reduced to three two-class classification problems in Fig. 16.2(b). Three identical MuSpiNNs using parameter values discussed earlier are used for the three modules. However, since the three modules are independent of each other, it is not necessary to do so because the suitability of the network architecture is, often, problem specific. Therefore, each module may be designed individually to maximize the computational efficiency and classification accuracy for the corresponding two-class classification problem.

One-fifth of the available data (30 training instances or samples) is used for training the network. The MuSpiNN dedicated to identifying a specific class is trained with 10 samples belonging to that class and 20 samples belonging to the other two classes (10 each). A sample Multi-SpikeProp convergence curve for each MuSpiNN module is shown in Fig. 16.3. It is observed that the three networks converge to an MSE of 0.2 in an average of 11, 60, and 28 epochs. The corresponding classification accuracies of the test data for the three classes are 100%, 94.5%, and 93.1%, which is higher than the 92.7%

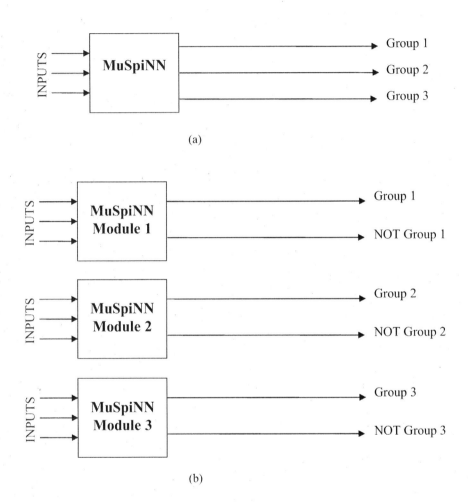

(a)

(b)

FIGURE 16.2
MuSpiNN architecture for (a) the original three-class classification problem
and (b) three two-class classification problems

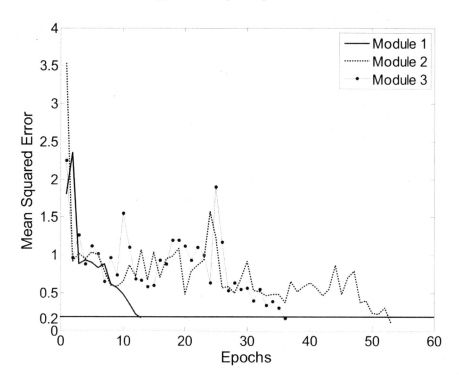

FIGURE 16.3
Sample Multi-SpikeProp convergence curves for MuSpiNN for the iris classification problem

classification accuracy obtained using the SNN with SpikeProp for the three-
class classification problem using similar network parameters and architectures
(discussed in Section 14.3). When half the available data (75 samples) are
used to train the network, similar classification accuracies are obtained but
the network needs to converge to a much lower MSE of 0.075.

16.5   EEG Classification Problem

The same feature space composed of the nine EEG features described in Sections 8.5 and 14.4.1 is employed in order to accurately classify the EEGs into the three linearly inseparable classes. The dataset consists of 300 samples. Similar to the iris problem, the three-class classification task is divided into three two-class classification tasks, each of which is solved by a separate dedicated MuSpiNN trained with Multi-SpikeProp. All parameter values are selected to be the same as those for the iris classification problem. The only difference is the number of input features, i.e., nine instead of four. The same population encoding scheme described for the iris classification problem is used with $M = 4$ input neurons for each of the nine input features plus a bias neuron, resulting in a total of $9M + 1 = 99 + 1 = 37$ input neurons.

One-tenth of the available dataset (30 training instances or data points) is used for training the network. The MuSpiNN dedicated to identifying a specific class is trained with 10 data points belonging to that class and 20 data points belonging to the other two classes (10 each). A sample Multi-SpikeProp convergence curve for each MuSpiNN module is shown in Fig. 16.4. It is observed that the three networks converge to an MSE of 0.2 in an average of 53, 36, and 78 epochs. The corresponding classification accuracies of the test data for the three classes are 90.7%, 91.5%, and 94.8%, which is significantly higher than the 82% classification accuracy obtained with SpikeProp for the three-class classification problem using similar network parameters and architectures (discussed in Section 14.4). When one-fifth of the available dataset (60 training instances) is used to train the network to a lower MSE of 0.075, similar classification accuracies are obtained for the first two classes but the

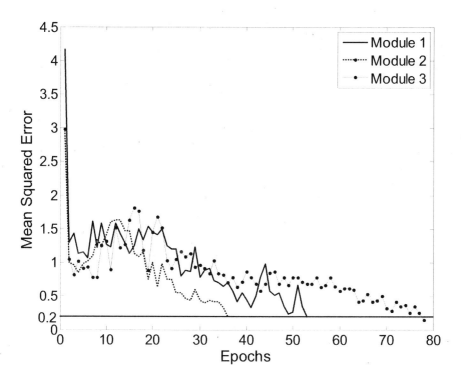

FIGURE 16.4
Sample Multi-SpikeProp convergence curves for MuSpiNN for the EEG classification problem

classification accuracy of the last class increases to 97.1%, a significant improvement, although at additional computational cost.

16.6 Discussion and Concluding Remarks

A new multi-spiking neural network (MuSpiNN) and a new learning algorithm, Multi-SpikeProp, for training the network have been presented in this chapter. The traditional BP-based supervised learning proposed by Rumelhart required a continuous and differentiable activation function for error backpropagation.

The lack of such a function for SNNs led to a strongly held belief that BP-based learning in SNNs is impossible. In 2002, Bohte et al. (2002a) developed a new learning rule that allowed such an SNN to learn based on adaptation in the timing of single spikes. Our novel network and learning algorithm advances that development to a more general case: BP-based learning in neural networks that can communicate via spike trains (i.e., multiple spikes instead of single spikes).

The performance of the network and learning algorithm was investigated using three different classification problems. It is found that MuSpiNN learns the XOR problem in twice the number of epochs compared with the single-spiking SNN model but requires only one-fourth the number of synapses. MuSpiNN and Multi-SpikeProp were also applied in a modular architecture to solve the three-class iris and EEG epilepsy and seizure detection problems, resulting in an increase in the classification accuracy compared with the single-spiking SNN and SpikeProp, especially in the case of the EEG problem. SNNs demonstrate great potential for solving complicated time-dependent pattern recognition problems defined by time series because of their inherent dynamic representation.

The application of MuSpiNN and Multi-SpikeProp demonstrates the training of a neural network that employs neurons based on an increased degree of biological plausibility compared to 1) traditional ANN neurons which simulate spike train communication with approximate implementations of rate encoding and 2) single-spiking SNN neurons (Bohte et al., 2002a). Although the MuSpiNN model can reconstruct temporal sequences of spikes as the network output, for the sake of simplicity we have restricted the output layer neuron to a single spike. An output spike train would require the selection of appropriate spike trains for representing the various classes and an appropriate error function. Research into the appropriateness of output spike trains and the error function will be another subject of future research toward harnessing the computational power of pulse encoding.

An obstacle to the use of more detailed neuronal models for such classification and pattern recognition tasks is imposed by the dynamics of the BP algorithm, which usually requires a single activation function (representing changes in membrane potential) for backpropagating the error term through the neuron. The detailed models are usually based on multiple differential equations that capture the behavior of different ion channels and currents that affect the membrane potential. It remains to be seen if error backpropagation is even mathematically possible in the face of such complexity. Alternatively, biologically plausible learning mechanisms such as Hebbian learning and spike time dependent plasticity (STDP) that have been used on such detailed models for demonstrating dynamics of small neuronal networks may need to be adapted for classification and complex pattern recognition tasks.

The biological plausibility of the backpropagation learning algorithm itself has been debated since its conception in the 1980s (Carpenter and Grossberg, 1987; Grossberg, 1988; Stork, 1989; Mazzoni et al., 1991). For the sake of discussion, consider a purely feedforward network where the flow of information is unidirectional. In such a network, implementations of BP similar to Rumelhart et al. (1986) by themselves are biologically unrealistic in the local sense that they do not directly model strengthening or weakening of a particular synapse based on the activity of the pre- and post-synaptic neurons for that synapse. However, the feedforward network and the BP learning algorithm together can arguably be an abstract representation of a biologically plausible system where the presynaptic neurons in the network get excitatory or inhibitory feedback based on the appropriateness of the final output from the network (i.e., the size of the error function). This abstraction models learning in a global sense because this feedback is an assumption that is external to the actual neural network. SpikeProp and Multi-SpikeProp have their origins in Rumelhart's BP concept, and therefore are only biologically plausible in the systemic sense. However, the Multi-SpikeProp learning algorithm results in the adjustment of individual spike times and could, in the future, be integrated with mechanisms such as STDP in order to increase the biological plausibility of the model.

A general shortcoming of the SNN models is the computational burden, which may be reduced by developing efficient learning algorithms. Additionally, the authors believe that the adaptive adjustment of the number of synapses discussed earlier will be especially effective in reducing the number of synaptic computations, without compromising the classification accuracy. Another source of computational effort is the input encoding that increases the number of features many fold. Novel methods of input encoding that do not increase the number of features and, at the same time, represent the input accurately must be explored. Another problem is the sensitivity of gradient descent-based learning algorithms to the initial state of the SNNs. This has been investigated and heuristic rules have been developed to ensure that the SNN training converges to a global minimum despite the highly uneven error surface (Bohte et al., 2002a; Moore, 2002; Xin and Embrechts, 2001; Booij and Nguyen, 2005; Silva and Ruano, 2005; McKennoch et al., 2006; Ghosh-Dastidar and Adeli, 2007). Although significant advances have been made in this aspect, the issue has not been fully resolved.

To avoid the sensitivity of the learning to the initial SNN state, non-gradient descent algorithms such as genetic algorithms (Hagras et al., 2004) and evolutionary strategies have been proposed (Belatreche et al., 2003; Pavlidis et al., 2005). Such algorithms have been used with considerable success for traditional ANNs and are still being investigated for use with SNNs. The learning algorithms presented by Hagras et al. (2004) and Pavlidis et al.

(2005) adapt only the weights to reach an optimum solution. The optimization objective usually involves minimization of a cost function (typically, a measure of error). Belatreche et al. (2003, 2007) presented an evolutionary strategy-based learning algorithm that adapts *both* synaptic weights and delays to reach the the optimization objective and reported classification performance comparable to Bohte et al.'s SpikeProp on the XOR and iris classification problems. Jin et al. (2007) presented a Pareto-based multi-objective genetic algorithm to simultaneously adjust the connectivity, weights, and delays of the SNN. The two optimization objectives of this model were to minimize: (1) the number of synaptic connections and (2) either the classification error or the root mean squared error. Despite some promising results, it should be noted that a known disadvantage of evolutionary algorithms is their prohibitively large computational cost for large problems.

SNNs discussed thus far have been feedforward and recurrent networks. These networks have been investigated mostly with regular and fully connected architectures. Such architectures are not common in biological neural networks. A biological network consists of many neurons and their synapses which form the physical network. In reality, however, only a small percentage of the neurons in the physical network contributes to the actual information processing. These neurons activate whereas neurons that do not contribute to the output remain silent. The network of activated neurons will be referred to as the *functional* network. Biological learning within such a network is a dynamic process. Silent neurons do not always remain silent and the activated neurons are not required to activate every time. The neurons that comprise the functional network often change based on the input.

This aspect of biological networks can be modeled using the *reservoir computing* paradigm. It is possible that biological plausibility was less of a motivation for this development than the search for more efficient supervised learning algorithms for recurrent ANNs (Paugam-Moisy and Bohte, 2009). Similar to the traditional recurrent ANNs, these SNNs have the traditional input and output neurons, some with recurrent connections. Between the input and the output neurons, instead of a hidden layer of neurons in the traditional sense, there is a set or *reservoir* of neurons. The architecture of the reservoir is random with sparse connections. Based on the input to the network, the input neurons activate a subset of neurons within the reservoir. This functional network, i.e., the spatial activation pattern, is read by the output neurons, which are called the *readout neurons*. It is important to note that in this model, supervised learning occurs *only* at the synaptic connections between the reservoir and the output layer, usually by means of algorithms such as linear regression. Two examples of reservoir computing, as presented for SNN models, are the Echo-State Network (Jaeger, 2001; Jaeger and Lukosevicius, 2007) and Liquid State Machines (Maass et al., 2002). For a comparison of these models the reader should refer to Schrauwen et al. (2007) and Paugam-Moisy and Bohte (2009).

Supervised Hebbian learning is another biologically plausible learning strategy that has been proposed for SNNs (Legenstein et al., 2005). In this strategy, similar to the other learning algorithms, the objective of the output neurons is to learn to fire at the desired spike times. However, in order to supervise the learning of the output neuron, the output neuron is constrained by additional inputs so that it fires only at specific times and not at others. This model was extended by Kasinski and Ponulak (2005) to a Liquid State Machine network. In this model, the supervision is not imposed directly on the input neuron. Rather, additional neurons outside the primary network receive the desired signals and influence the synaptic learning in the network. This model is dubbed Remote Supervision Method because the task of supervision is performed by the external neurons. This model was shown to have a high learning ability and accuracy (Kasinski and Ponulak, 2005).

Researchers have demonstrated that SNNs have significant potential for solving complicated time-dependent pattern recognition problems because of dynamic representation inherent in spiking neurons. Moreover, SNNs have been theoretically shown to have the ability to approximate any continuous function (Maass, 1997a). The addition of the temporal dimension for information encoding has the potential to result in compact representations of large neural networks, another advantage for SNNs. Despite the interest in supervised learning algorithms for SNNs, their widespread acceptance and development are currently limited by the excessive computing times required for training (Ghosh-Dastidar and Adeli, 2007). This is a problem even with algorithms that are not based on gradient descent, especially when the network size becomes large. This shortcoming may be eliminated in the near future due to advances in computing technology in general as well as the development of more efficient and accurate SNN learning algorithms. Some of these algorithms have been discussed in this chapter. Combinations of such novel strategies along with improved understanding of biological information processing will contribute significantly to the development of SNNs as the next generation neural networks.

17

The Future

This book presented a novel approach and powerful computational algorithms for automated EEG-based diagnosis of neurological disorders. The primary application area was epilepsy diagnosis and seizure detection. One reason for the focus on epilepsy was the availability of EEG data. Some preliminary results were also presented for diagnosis of Alzheimer's disease (AD). The authors could not present a complete method for AD diagnosis because sufficient data were not available to test the models under development.

The book, however, presents a general approach and methodology that the authors believe will be the wave of the future and an important tool in the practice of neurology. The methodology is general and can be adapted and applied for diagnosis of other neurological disorders. In fact, while the authors were writing this book, the senior author and an associate developed a methodology for EEG-based diagnosis of attention-deficit/hyperactivity disorder (ADHD) through adroit integration of nonlinear science, wavelets, and neural networks. The selected nonlinear features are generalized synchronizations known as synchronization likelihoods. The methodology has been applied to EEG data obtained from 47 ADHD and 7 control individuals with eyes closed. Using a radial basis function neural network classifier the methodology yielded a high accuracy of 96.5% for diagnosis of the ADHD. The senior author and his associates are expanding the work for early diagnosis of AD as well as diagnosis of autism.

From a physiologic perspective, the temporal evolution of the EEG represents one aspect of brain dynamics. Spatial changes (across various key regions in the brain) are equally important and should be investigated to obtain a comprehensive understanding of the spatio-temporal dynamics in the brain. Such analyses have the potential to yield additional quantifiable markers of abnormality in neurological disorders, as was observed in our preliminary investigation of AD. For complex phenomena, a single modality of investigation (such as imaging study or EEG study) may also not be sufficient and multi-modal spatio-temporal studies may be more effective. For instance, a better understanding of the changes in brain activity during the performance of various mental tasks (as obtained from fMRI studies) could identify linked areas of the brain in which to focus EEG studies. In the future, as the field matures, the effectiveness of a combination of various modalities should be investigated for improving the efficacy of detection and diagnosis models.

A second contribution of the book is presenting and advancing spiking neural networks (SNNs) as the foundation of a more realistic and plausible third generation neural network. In addition to their application as classifiers presented in this book, SNNs can be used for modeling populations of neurons and their interactions, which could provide explanations for dynamics in the cerebral cortex that cannot be modeled by current mathematical models. It is hoped the research in this area will advance in the coming years, resulting in more powerful computational neural network models not only for diagnosis of neurological disorders but also for the solution of many other complex and intractable time-dependent pattern recognition and prediction phenomena.

Bibliography

Abasolo, D., Hornero, R., Espino, P., Poza, J., Sanchez, C. I., and de la Rosa, R. (2005). Analysis of regularity in the EEG background activity of Alzheimer's disease patients with approximate entropy. *Clinical Neurophysiology*, 116(8):1826–1834.

Abbott, L. F. and Kepler, T. B. (1990). Model neurons: From Hodgkin-Huxley to Hopfield. In Garrido, L., editor, *Statistical Mechanics of Neural Networks*. Springer, Berlin.

Abry, P. (1997). *Ondelettes et turbulence. Multirésolutions, Algorithmes de Décomposition, Invariance D'échelles*. Diderot Editeur, Paris, France.

Adeli, H. (2001). Neural networks in civil engineering: 1989-2000. *Computer-Aided Civil and Infrastructure Engineering*, 16(2):126–142.

Adeli, H. and Ghosh-Dastidar, S. (2004). Mesoscopic-wavelet freeway work zone flow and congestion feature extraction model. *Journal of Transportation Engineering*, 130(1):94–103.

Adeli, H., Ghosh-Dastidar, S., and Dadmehr, N. (2005a). Alzheimer's disease and models of computation: Imaging, classification, and neural models. *Journal of Alzheimer's Disease*, 7(3):187–199.

Adeli, H., Ghosh-Dastidar, S., and Dadmehr, N. (2005b). Alzheimer's disease: Models of computation and analysis of EEGs. *Clinical EEG and Neuroscience*, 36(3):131–140.

Adeli, H., Ghosh-Dastidar, S., and Dadmehr, N. (2007). A wavelet-chaos methodology for analysis of EEGs and EEG sub-bands to detect seizure and epilepsy. *IEEE Transactions on Biomedical Engineering*, 54(2):205–211.

Adeli, H. and Hung, S. L. (1994). An adaptive conjugate gradient learning algorithm for effective training of multilayer neural networks. *Applied Mathematics and Computation*, 62(1):81–102.

Adeli, H. and Hung, S. L. (1995). *Machine Learning - Neural Networks, Genetic Algorithms, and Fuzzy Sets*. John Wiley and Sons, NY.

Adeli, H. and Jiang, X. (2003). Neuro-fuzzy logic model for freeway work zone capacity estimation. *Journal of Transportation Engineering*, 129(5):484–493.

Adeli, H. and Jiang, X. (2006). Dynamic fuzzy wavelet neural network model for structural system identification. *Journal of Structural Engineering*, 132(1):102–111.

Adeli, H. and Karim, A. (1997). Neural dynamics model for optimization of cold-formed steel beams. *Journal of Structural Engineering*, 123(11):1535–1543.

Adeli, H. and Karim, A. (2000). Fuzzy-wavelet RBFNN model for freeway incident detection. *Journal of Transportation Engineering*, 126(6):464–471.

Adeli, H. and Karim, A. (2005). *Wavelets in Intelligent Transportation Systems*. John Wiley and Sons, Hoboken, NJ.

Adeli, H. and Park, H. S. (1995a). Counter propagation neural network in structural engineering. *Journal of Structural Engineering*, 121(8):1205–1212.

Adeli, H. and Park, H. S. (1995b). Optimization of space structures by neural dynamics. *Neural Networks*, 8(5):769–781.

Adeli, H. and Park, H. S. (1998). *Neurocomputing for Design Automation*. CRC Press, Boca Raton, FL.

Adeli, H. and Samant, A. (2000). An adaptive conjugate gradient neural network - wavelet model for traffic incident detection. *Computer-Aided Civil and Infrastructure Engineering*, 15(4):251–260.

Adeli, H. and Wu, M. (1998). Regularization neural network for construction cost estimation. *Journal of Construction Engineering and Management*, 124(1):18–24.

Adeli, H., Zhou, Z., and Dadmehr, N. (2003). Analysis of EEG records in an epileptic patient using wavelet transform. *Journal of Neuroscience Methods*, 123(1):69–87.

Ademoglu, A., Micheli-Tzanakou, E., and Istefanopulos, Y. (1997). Analysis of pattern reversal visual evoked potentials (PRVEPs) by spline wavelets. *IEEE Transactions on Biomedical Engineering*, 44:881–890.

Aftanas, L. I. and Golocheikine, S. A. (2002). Non-linear dynamic complexity of the human EEG during meditation. *Neuroscience Letters*, 330(2):143–146.

American Clinical Neurophysiology Society (2006). Guidelines for standard electrode position nomenclature. ACNS, Bloomfield, CT. Available: https://www.acns.org.

American Psychiatric Association (1994). *Diagnostic and Statistical Manual of Mental Disorders (DSM-IV)*. APA, Washington, D.C., 4th edition.

Anderer, P., Saletu, B., Kloppel, B., Semlitsch, H. V., and Werner, H. (1994). Discrimination between demented patients and normals based on topographic EEG slow wave activity: Comparison between Z statistics, discriminant analysis and artificial neural network classifiers. *Electroencephalography and Clinical Neurophysiology*, 91(2):108–117.

Andersen, A. H., Zhang, Z., Avison, M. J., and Gash, D. M. (2002). Automated segmentation of multispectral brain MR images. *Journal of Neuroscience Methods*, 122:13–23.

Andrzejak, R. G., Lehnertz, K., Rieke, C., Mormann, F., David, P., and Elger, C. E. (2001). Indications of non-linear deterministic and finite dimensional structures in time series of brain electrical activity: Dependence on recording region and brain state. *Physical Review E*, 64(6):1–8(061907).

Anokhin, A. P., Birbaumer, N., Lutzenberger, W., Nikolaev, A., and Vogel, F. (1996). Age increases brain complexity. *Electroencephalography and Clinical Neurophysiology*, 99:63–68.

Ashburner, J. and Friston, K. J. (2000). Voxel-based morphometry - the methods. *NeuroImage*, 11:805–821.

Ashburner, J., Hutton, C., Frackowiak, R., Johnsrude, I., Price, C., and Friston, K. (1998). Identifying global anatomical differences: Deformation-based morphometry. *Human Brain Mapping*, 6:348–357.

Avila, J., Lim, F., Moreno, F., Belmonte, C., and Cuello, A. C. (2002). Tau function and dysfunction in neurons: Its role in neurodegenerative disorders. *Molecular Neurobiology*, 25:213–231.

Baloyannis, S. J., Costa, V., and Michmizos, D. (2004). Mitochondrial alterations in Alzheimer's disease. *American Journal of Alzheimer's Disease and Other Dementias*, 19:89–93.

Baron, J. C., Chetelat, G., Desgranges, B., Perchey, G., Landeau, B., de la Sayette, V., and Eustache, F. (2001). In vivo mapping of gray matter loss with voxel-based morphometry in mild Alzheimer's disease. *NeuroImage*, 14:298–309.

Barra, V. and Boire, J. Y. (2000). Tissue segmentation on MR images of the brain by possibilistic clustering on a 3D wavelet representation. *Journal of Magnetic Resonance Imaging*, 11:267–278.

Belatreche, A., Maguire, L. P., and McGinnity, M. (2007). Advances in design and application of spiking neural networks. *Soft Computing - A Fusion of Foundations, Methodologies and Applications*, 11:239–248.

Belatreche, A., Maguire, L. P., McGinnity, M., and Wu, Q. X. (2003). A method for supervised training of spiking neural networks. In *Proceedings* of the IEEE Conference Cybernetics Intelligence Challenges and Advances, page 3944, Reading, UK.

Bellman, R. (1961). *Adaptive Control Processes: A Guided Tour*. Princeton University Press, Princeton, NJ.

Bennys, K., Rondouin, G., Vergnes, C., and Touchon, J. (2001). Diagnostic value of quantitative EEG in Alzheimer's disease. *Neurophysiologie Clinique (Clinical Neurophysiology)*, 31:153–160.

Benvenuto, J., Jin, Y., Casale, M., Lynch, G., and Granger, R. (2002). Identification of diagnostic evoked response potential segments in Alzheimer's disease. *Experimental Neurology*, 176:269–276.

Berendse, H. W., Verbunt, J. P. A., Scheltens, P., van Dijk, B. W., and Jonkman, E. J. (2000). Magnetoencephalographic analysis of cortical activity in Alzheimer's disease: A pilot study. *Clinical Neurophysiology*, 111:604–612.

Besthorn, C., Frstl, H., Geiger-Kabisch, C., Sattel, H., Gasser, T., and Schreiter-Gasser, U. (1994). EEG coherence in Alzheimer's disease. *Electroencephalography and Clinical Neurophysiology*, 90:242–245.

Besthorn, C., Sattel, H., Geiger-Kabisch, C., Zerfass, R., and Frstl, H. (1995). Parameters of EEG dimensional complexity in Alzheimer's disease. *Electroencephalography and Clinical Neurophysiology*, 95:84–89.

Besthorn, C., Zerfass, R., Geiger-Kabisch, C., Sattel, H., Daniel, S., Schreiter-Gasser, U., and Frstl, H. (1997). Discrimination of Alzheimer's disease and normal aging by EEG data. *Electroencephalography and Clinical Neurophysiology*, 103(2):241–248.

Block, W., Traber, F., Flacke, S., Jessen, F., Pohl, C., and Schild, H. (2002). In-vivo proton MR-spectroscopy of the human brain: Assessment of N-acetylaspartate (NAA) reduction as a marker for neurodegeneration. *Amino Acids*, 23:317–323.

Bohte, S. M., Kok, J. N., and La Poutr, H. (2002a). Unsupervised classification in a layered network of spiking neurons. *IEEE Transactions on Neural Networks*, 13(2):426–435.

Bohte, S. M., Kok, J. N., and La Poutr, J. A. (2002b). Error-backpropagation in temporally encoded networks of spiking neurons. *Neurocomputing*, 48(1-4):17–37.

Booij, O. and Nguyen, H. T. (2005). A gradient descent rule for multiple spiking neurons emitting multiple spikes. *Information Processing Letters*, 95(6):552–558.

Borovkova, S., Burton, R., and Dehling, H. (1999). Consistency of the Takens estimator for the correlation dimension. *The Annals of Applied Probability*, 9(2):376–390.

Bose, N. K. and Liang, P. (1996). *Neural Network Fundamentals with Graphs, Algorithms, and Applications*. McGraw-Hill, NY.

Bottino, C. M., Castro, C. C., Gomes, R. L., Buchpiguel, C. A., Marchetti, R. L., and Neto, M. R. (2002). Volumetric MRI measurements can differentiate Alzheimer's disease, mild cognitive impairment, and normal aging. *International Psychogeriatrics*, 14:59–72.

Bozzao, A., Floris, R., Baviera, M. E., Apruzzese, A., and Simonetti, G. (2001). Diffusion and perfusion MR imaging in cases of Alzheimer's disease: Correlations with cortical atrophy and lesion load. *American Journal of Neuroradiology*, 22:1030–1036.

Budinger, T. F. (1996). Neuroimaging applications for the study of Alzheimer's disease. In Khachaturian, Z. S. and Radebaugh, T. S., editors, *Alzheimer's Disease: Cause(s), Diagnosis, Treatment and Care*, pages 145–174. CRC Press, Boca Raton, FL.

Bullmore, E., Brammer, M., Alarcon, G., and Binnie, C. (1992). A new technique for fractal analysis applied to human, intracerebrally recorded, ictal electroencephalographic signals. In Sataloff, R. T. and Hawkshaw, M., editors, *Chaos in Medicine: Source Readings*, pages 295–298. Singular Publishing Group, San Diego, CA.

Burgess, A. P., Rehman, J., and Williams, J. D. (2003). Changes in neural complexity during the perception of 3D images using random dot stereograms. *International Journal of Psychophysiology*, 48:35–42.

Burrus, C. S., Gopinath, R., and Guo, H. (1998). *Introduction to Wavelets and Wavelet Transforms: A Primer*. Prentice Hall, New Jersey.

Burton, E. J., Karas, G., Paling, S. M., Barber, R., Williams, E. D., Ballard, C. G., McKeith, I. G., Scheltens, P., Barkhof, F., and O'Brien, J. T. (2002). Patterns of cerebral atrophy in dementia with Lewy bodies using voxel-based morphometry. *NeuroImage*, 17:618–630.

Cao, L. (1997). Practical method for determining the minimum embedding dimension of a scalar time series. *Physica D*, 110(1-2):43–50.

Carpenter, G. A. and Grossberg, S. (1987). A massively parallel architecture for a self-organizing neural pattern recognition machine. *Computer Vision, Graphics, and Image Processing*, 37:54–115.

Chui, C. K. (1992). *An Introduction to Wavelets*. Academic Press, San Diego, CA.

Coben, L. A., Danziger, W., and Storandt, M. (1985). A longitudinal EEG study of mild senile dementia of Alzheimer type: Changes at 1 year and at 2.5 years. *Electroencephalography and Clinical Neurophysiology*, 61:101–112.

Cook, I. A. and Leuchter, A. F. (1996). Synaptic dysfunction in Alzheimer's disease: Clinical assessment using quantitative EEG. *Behavioral Brain Research*, 78:15–23.

Cruz, L., Urbanc, B., Buldyrev, S. V., Christie, R., Gomez-Isla, T., Havlin, S., McNamara, M., Stanley, H. E., and Hyman, B. T. (1997). Aggregation and disaggregation of senile plaques in Alzheimer disease. *Proceedings of the National Academy of Sciences*, 94:7612–7616.

Dahlbeck, S. W., McCluney, K. W., Yeakley, J. W., Fenstermacher, M. J., Bonmati, C., and Van Horn, G. (1991). The interuncal distance: A new RM measurement for the hippocampal atrophy in Alzheimer's disease. *American Journal of Neuroradiology*, 12:931–932.

Daubechies, I. (1988). Orthonormal bases of compactly supported wavelets. *Communications on Pure and Applied Mathematics*, 41:909–996.

Daubechies, I. (1992). *Ten Lectures on Wavelets*. Society for Industrial and Applied Mathematics, Philadelphia, PA.

de Courten-Myers, G. M. (2004). Cerebral amyloid angiopathy and Alzheimer's disease. *Neurobiology of Aging*, 25:603–604.

deFigueiredo, R. J., Shankle, W. R., Maccato, A., Dick, M. B., Mundkur, P., Mena, I., and Cotman, C. (1995). Neural-network-based classification of cognitively normal, demented, Alzheimer's disease and vascular dementia from single photon emission with computed tomography image data from brain. *Proceedings of the National Academy of Sciences*, 92:5530–5534.

Deng, X., Li, K., and Liu, S. (1999). Preliminary study on application of artificial neural network to the diagnosis of Alzheimer's disease with magnetic resonance imaging. *Chinese Medical Journal (Beijing)*, 112:232–237.

Desphande, N. A., Gao, F. Q., Bakshi, S. N., Leibovitch, F. S., and Black, S. E. (2004). Simple linear and area MR measurements can help distinguish between Alzheimer's disease, frontotemporal dementia, and normal aging: The Sunnybrook dementia study. *Brain and Cognition*, 54:165–166.

Devinsky, O. (2004). Diagnosis and treatment of temporal lobe epilepsy. *Reviews in Neurological Diseases*, 1(1):2–9.

Dharia, A. and Adeli, H. (2003). Neural network model for rapid forecasting of freeway link travel time. *Engineering Applications of Artificial Intelligence*, 16(7-8):607–613.

Dierks, T., Perisic, I., Frolich, L., Ihl, R., and Maurer, K. (1991). Topography of quantitative electroencephalogram in dementia of the Alzheimer type: Relation to severity of dementia. *Psychiatry Research*, 40:181–194.

Dillon, W. R. and Goldstein, M. (1984). *Multivariate Analysis: Methods and Applications*. John Wiley and Sons, New York.

Doraiswamy, P. M., McDonald, W. M., Patterson, L., Husain, M. M., Figiel, G. S., Boyko, O. B., and Krishnan, K. R. (1993). Interuncal distance as a measure of hippocampal atrophy: Normative data on axial MR imaging. *American Journal of Neuroradiology*, 14:141–143.

Duch, W. (2000). Therapeutic implications of computer models of brain activity for Alzheimer disease. *Journal of Medical Informatics and Technologies*, 5:27–34.

Duffy, F. H., Albert, M. S., and McAnulty, G. (1984). Brain electrical activity in patients with presenile and senile dementia of the Alzheimer type. *Annals of Neurology*, 16:439–448.

Dunkin, J. J., Osato, S., and Leuchter, A. F. (1995). Relationships between EEG coherence and neuropsychological tests in dementia. *Clinical Electroencephalography*, 26:47–59.

Durka, P. J. (2003). From wavelets to adaptive approximations: Time-frequency parameterization of EEG. *BioMedical Engineering OnLine*, 2(1):1(1–30).

Dvorak, I. (1990). Takens versus multichannel reconstruction in EEG correlation exponent estimates. *Physics Letters A*, 151:225–233.

Early, B., Escalona, P. R., Boyko, O. B., Doraiswamy, P. M., Axelson, D. A., Patterson, L., McDonald, W. M., and Krishnan, K. R. (1993). Interuncal distance measurements in healthy volunteers and in patients with Alzheimer disease. *American Journal of Neuroradiology*, 14:907–910.

Eckert, A., Keil, U., Marques, C. A., Bonert, A., Frey, C., Schussel, K., and Muller, W. E. (2003). Mitochondrial dysfunction, apoptotic cell death, and Alzheimer's disease. *Biochemical Pharmacology*, 66:1627–1634.

Efremova, T. M. and Kulikov, M. A. (2002). Chaotic component of human high-frequency EEG in the state of quiet wakefulness. *Zhurnal Vysshei Nervnoi Deiatelnosti Imeni I P Pavlova*, 52(3):283–291.

Ehlers, C. L., Havstad, J., Pritchard, D., and Theiler, J. (1998). Low doses of ethanol reduce evidence for nonlinear structure in brain activity. *Journal of Neuroscience*, 18:7474–7486.

El Fakhri, G., Moore, S. C., Maksud, P., Aurengo, A., and Kijewski, M. (2001). Absolute activity quantitation in simultaneous 123I/99mtc brain SPECT. *Journal of Nuclear Medicine*, 42:300–308.

Elger, C. E. and Lehnertz, K. (1994). Ictogenesis and chaos. In Wolf, P., editor, *Epileptic Seizures and Syndromes*, pages 547–552. Libbey, London, U.K.

Elger, C. E. and Lehnertz, K. (1998). Seizure prediction by non-linear time series analysis of brain electrical activity. *European Journal of Neuroscience*, 10(2):786–789.

Erkinjuntti, T., Lee, D. H., Gao, F., Steenhuis, R., Eliasziw, M., Fry, R., Merskey, H., and Hachinski, V. C. (1993). Temporal lobe atrophy on magnetic resonance imaging in the diagnosis of early Alzheimer's disease. *Archives of Neurology*, 50:305–310.

Ermentrout, G. B. (1996). Type I membranes, phase resetting curves, and synchrony. *Neural Computation*, 8:979–1001.

Ermentrout, G. B. and Kopell, N. (1986). Parabolic bursting in an excitable system coupled with a slow oscillation. *SIAM Journal on Applied Mathematics*, 46:233–253.

Escudero, J., Abasolo, D., Hornero, R., Espino, P., and Lopez, M. (2006). Analysis of electroencephalograms in Alzheimer's disease patients with multiscale entropy. *Physiological Measurement*, 27(11):1091–1106.

Evans, P. H. (1993). Free radicals in brain metabolism and pathology. *British Medical Bulletin*, 49:577–587.

Fahlman, S. E. (1988). Faster-learning variations of back-propagation: An empirical study. In *Proceedings of the 1988 Connectionist Models Summer School*, pages 38–51, San Mateo, CA. Morgan Kaufmann.

Fell, J., Roschke, J., and Schaffner, C. (1996). Surrogate data analysis of sleep electroencephalograms reveals evidence for nonlinearity. *Biological Cybernetics*, 75:85–92.

Ferri, R., Chiaramonti, R., Elia, M., Musumeci, S. A., Ragazzoni, A., and Stam, C. J. (2003). Non-linear EEG analysis during sleep in premature and full-term newborns. *Clinical Neurophysiology*, 114(7):1176–1180.

Ferri, R., Parrino, L., Smerieri, A., Terzano, M. G., Elia, M., Musumeci, S. A., Pettinato, S., and Stam, C. J. (2002). Non-linear EEG measures during sleep: Effects of the different sleep stages and cyclic alternating pattern. *International Journal of Psychophysiology*, 43(3):273–286.

Ferri, R., Pettinato, S., Alicata, F., Del Gracco, S., Elia, M., and Musumeci, S. (1998). Correlation dimension of EEG slow-wave activity during sleep in children and young adults. *Electroencephalography and Clinical Neurophysiology*, 106(5):424–428.

Finkel, L. H. (2000). Neuroengineering models of brain disease. *Annual Review of Biomedical Engineering*, 2:577–606.

Fisch, B. J. (1999). *EEG Primer: Basic Principles of Digital and Analog EEG.* Elsevier Science Publishers, Amsterdam, Netherlands, 3rd edition.

Fisher, R. A. (1936). The use of multiple measurements in taxonomic problems. *Annals of Eugenics*, 7:179–188.

Fox, N. C., Crum, W. R., Scahill, R. I., Stevens, J. M., Janssen, J. C., and Rossor, M. N. (2001). Imaging of onset and progression of Alzheimer's disease with voxel-compression mapping of serial magnetic resonance images. *The Lancet*, 358:201–205.

Frisoni, G. B., Beltramello, A., Weiss, C., Geroldi, C., Bianchetti, A., and Trabucchi, M. (1996). Linear measures of atrophy in mild Alzheimer disease. *American Journal of Neuroradiology*, 17:913–923.

Frisoni, G. B., Geroldi, C., Beltramello, A., Bianchetti, A., Binetti, G., Bordiga, G., DeCarli, C., Laakso, M. P., Soininen, H., Testa, C., Zanetti, O., and Trabucchi, M. (2002). Radial width of the temporal horn: A sensitive measure in Alzheimer disease. *American Journal of Neuroradiology*, 23:35–47.

Fukunaga, K. (1990). *Introduction to Statistical Pattern Recognition.* Academic Press, New York, 2nd edition.

Gabor, D. (1946). Theory of communication. *Journal of Institute of Electrical Engineering*, 93(3):429–457.

Gao, F. Q., Black, S. E., Leibovitch, F. S., Callen, D. J., Lobaugh, N. J., and Szalai, J. P. (2003). A reliable MR measurement of medial temporal lobe width from the Sunnybrook dementia study. *Neurobiology of Aging*, 24:49–56.

Gerstner, W. (1995). Time structure of the activity in neural network models. *Physical Review E*, 51:738–758.

Gerstner, W., Kempter, R., van Hemmen, J., and Wagner, H. (1996). A neuronal learning rule for sub-millisecond temporal coding. *Nature*, 383:76–78.

Gerstner, W. and Kistler, W. M. (2002). *Spiking Neuron Models. Single Neurons, Populations, Plasticity.* Cambridge University Press, New York.

Ghosh-Dastidar, S. and Adeli, H. (2003). Wavelet-clustering-neural network model for freeway incident detection. *Computer-Aided Civil and Infrastructure Engineering*, 18(5):325–338.

Ghosh-Dastidar, S. and Adeli, H. (2006). Neural network-wavelet microsimulation model for delay and queue length estimation at freeway work zones. *Journal of Transportation Engineering*, 132(4):331–341.

Ghosh-Dastidar, S. and Adeli, H. (2007). Improved spiking neural networks for EEG classification and epilepsy and seizure detection. *Integrated Computer-Aided Engineering*, 14(3):187–212.

Ghosh-Dastidar, S. and Adeli, H. (2009a). A new supervised learning algorithm for multiple spiking neural networks with application in epilepsy and seizure detection. *Neural Networks*, 22(10):1419–1431.

Ghosh-Dastidar, S. and Adeli, H. (2009b). Spiking neural networks. *International Journal of Neural Systems*, 19(4):295–308.

Ghosh-Dastidar, S., Adeli, H., and Dadmehr, N. (2007). Mixed-band wavelet-chaos-neural network methodology for epilepsy and epileptic seizure detection. *IEEE Transactions on Biomedical Engineering*, 54(9):1545–1551.

Ghosh-Dastidar, S., Adeli, H., and Dadmehr, N. (2008). Principal component analysis-enhanced cosine radial basis function neural network for robust epilepsy and seizure detection. *IEEE Transactions on Biomedical Engineering*, 55(2):512–518.

Goswami, J. C. and Chan, A. K. (1999). *Fundamentals of Wavelets*. John Wiley and Sons, New York.

Gotman, J. (1990). The use of computers in analysis and display of EEG and evoked potentials. In Daly, D. and Pedley, T., editors, *Current Practice of Clinical Electroencephalography*, pages 51–83. Raven Press, New York, 2nd edition.

Gotman, J. (1999). Inpatient/outpatient EEG monitoring. *Journal of Clinical Neurophysiology*, 16(2):130–140.

Grau, V., Mewes, A. U., Alcaniz, M., Kikinis, R., and Warfield, S. K. (2004). Improved watershed transform for medical image segmentation using prior information. *IEEE Transactions on Medical Imaging*, 23:447–458.

Grossberg, S. (1982). *Studies of Mind and Brain*. Reidel Press, Boston, MA.

Grossberg, S. (1988). Competitive learning: From interactive activation to adaptive resonance. In Waltz, D. and Feldman, J. A., editors, *Connectionist Models and Their Implications: Readings from Cognitive Science*, pages 243–283. Ablex, Norwood, NJ.

Grossberg, S. and Versace, M. (2008). Spikes, synchrony, and attentive learning by laminar thalamocortical circuits. *Brain Research*, 1218C:278–312.

Grossman, A. and Morlet, J. (1984). Decomposition of Hardy functions into square integrable wavelets of constant shape. *SIAM Journal on Mathematical Analysis*, 15:723–736.

Gu, F. J., Meng, X., Shen, E. H., and Cai, Z. J. (2003). Can we measure consciousness with EEG complexities? *International Journal of Bifurcation and Chaos*, 13:733–742.

Gueguen, B., Derouesne, C., Bourdel, M. C., Guillou, S., Landre, E., Gaches, J., Hossard, H., Ancri, D., and Mann, M. (1991). Quantified EEG in the diagnosis of Alzheimer's type dementia. *Clinical Neurophysiology*, 21:357–371.

Gueorguieva, N., Valova, I., and Georgiev, G. (2006). Learning and data clustering with an RBF-based spiking neuron network. *Journal of Experimental and Theoretical Artificial Intelligence*, 18(1):73–86.

Güler, N. F., Übeyli, E., and Güler, I. (2005). Recurrent neural networks employing Lyapunov exponents for EEG signals classification. *Expert Systems with Applications*, 29(3):506–514.

Gutkin, B., Pinto, D., and Ermentrout, B. (2003). Mathematical neuroscience: from neurons to circuits to systems. *Journal of Physiology - Paris*, 97:209–219.

Gutkin, B. S. and Ermentrout, G. B. (1998). Dynamics of membrane excitability determine interspike interval variability: a link between spike generation mechanisms and cortical spike train statistics. *Neural Computation*, 10(5):1047–1065.

Hagan, M. T., Demuth, H. B., and Beale, M. (1996). *Neural Network Design*. PWS Publishing Company, Boston, MA.

Haglund, M., Sjobeck, M., and Englund, E. (2004). Severe cerebral amyloid angiopathy characterizes an underestimated variant of vascular dementia. *Dementia and Geriatric Cognitive Disorders*, 18:132–137.

Hagras, H., Pounds-Cornish, A., and Colley, M. (2004). Evolving spiking neural network controllers for autonomous robots. In *Proceedings of the IEEE International Conference on Robotics and Automation*, volume 5, pages 4620–4626.

Hamilton, D., O'Mahony, D., Coffey, J., Murphy, J., O'Hare, N., Freyne, P., Walsh, B., and Coakley, D. (1997). Classification of mild Alzheimer's disease by artificial neural network analysis of SPET data. *Nuclear Medicine Communications*, 18:805–810.

Hampel, H., Teipel, S. J., Bayer, W., Alexander, G. E., Schwarz, R., Schapiro, M. B., Rapoport, S. I., and Moller, H. J. (2002). Age transformation of combined hippocampus and amygdala volume improves diagnostic accuracy in Alzheimer's disease. *Journal of the Neurological Sciences*, 194:15–19.

Hanyu, H., Shindo, H., Kakizaki, D., Abe, K., Iwamoto, T., and Takasaki, M. (1997). Increased water diffusion in cerebral white matter in Alzheimer's disease. *Gerontology*, 43:343–351.

Hasselmo, M. E. (1994). Runaway synaptic modification in models of cortex: Implications for Alzheimer's disease. *Neural Networks*, 7:13–40.

Hasselmo, M. E. (1995). Neuromodulation and cortical function: Modeling the physiological basis of behavior. *Behavioral Brain Research*, 67:1–27.

Hasselmo, M. E. and McClelland, J. L. (1999). Neural models of memory. *Current Opinion in Neurobiology*, 9:184–188.

Hecht-Nielsen, R. (1988). Application of counterpropagation networks. *Neural Networks*, 1(2):131–139.

Henderson, G. T., Ifeachor, E. C., Wimalartna, H. S. K., Allen, E. M., and Hudson, N. R. (2002). Electroencephalogram-based methods for routine detection of dementia. In Cerutti, S., Akay, M., Mainardi, L. T., and Zywietz, C., editors, *IFMBE-IMIA Proceedings of the 4th International Workshop on Biosignal Interpretation, Como, Italy*, pages 319–322, Edmonton, AB, Canada. IFMBE.

Heyman, A. (1996). Heterogeneity of Alzheimer's disease. In Khachaturian, Z. S. and Radebaugh, T. S., editors, *Alzheimer's Disease: Cause(s), Diagnosis, Treatment and Care*, pages 105–107. CRC Press, Boca Raton, FL.

Higdon, R., Foster, N. L., Koeppe, R. A., DeCarli, C. S., Jagust, W. J., Clark, C. M., Barbas, N. R., Arnold, S. E., Turner, R. S., Heidebrink, J. L., and Minoshima, S. (2004). A comparison of classification methods for differentiating fronto-temporal dementia from Alzheimer's disease using FDG-PET imaging. *Statistics in Medicine*, 23:315–326.

Hilborn, R. C. (2001). *Chaos and Nonlinear Dynamics: An Introduction for Scientists and Engineers*. Oxford University Press, Oxford, UK.

Hill, D. L., Batchelor, P. G., Holden, M., and Hawkes, D. J. (2001). Medical image registration. *Physics in Medicine and Biology*, 46:R1–45.

Hille, B. (1992). *Ionic Channels of Excitable Membranes*. Sinauer Associates, Sunderland, MA, 2nd edition.

Hively, L. M., Gailey, P. C., and Protopopescu, V. A. (1999). Detecting dynamical change in nonlinear time series. *Physics Letters A*, 258(2-3):103–114.

Hodgkin, A. L. and Huxley, A. F. (1952). A quantitative description of ion currents and its applications to conduction and excitation in nerve membranes. *Journal of Physiology*, 117:500–544.

Hopfield, J. (1995). Pattern recognition computation using action potential timing for stimulus representation. *Nature*, 376:33–36.

Hopfield, J. J. (1982). Neural networks and physical systems with emergent collective computational properties. *Proceedings of the National Academy of Sciences*, 79:2554–2558.

Hoppensteadt, F. C. and Izhikevich, E. M. (1997). *Weakly Connected Neural Networks*. Springer-Verlag, New York.

Horn, D., Levy, N., and Ruppin, E. (1996). Neuronal-based synaptic compensation: A computational study in Alzheimer's disease. *Neural Computation*, 8:1227–1243.

Horn, D., Ruppin, E., Usher, M., and Herrmann, M. (1993). Neural network modeling of memory deterioration in Alzheimer's disease. *Neural Computation*, 5:736–749.

Howlett, R. J. and Jain, L. C. (2001a). *Radial Basis Function Networks 1: Recent Developments in Theory and Applications*. Springer, Berlin, Germany.

Howlett, R. J. and Jain, L. C. (2001b). *Radial Basis Function Networks 2: New Advances in Design*. Springer, Berlin, Germany.

Hoyer, S. (1996). Oxidative metabolism deficiencies in brains of patients with Alzheimer's disease. *Acta Neurologica Scandinavica, Supplementum*, 165:18–24.

Huang, C., Wahlund, L., Dierks, T., Julin, P., Winblad, B., and Jelic, V. (2000). Discrimination of Alzheimer's disease and mild cognitive impairment by equivalent EEG sources: A cross-sectional and longitudinal study. *Clinical Neurophysiology*, 111:1961–1967.

Hubbard, B. (1998). *The World According to Wavelets: The Story of a Mathematical Technique in the Making*. A. K. Peters, Wellesley, MA, 2nd edition.

Huber, M. T., Braun, H. A., and Krieg, J. C. (1999). Consequences of deterministic and random dynamics for the course of affective disorders. *Biological Psychiatry*, 46(2):256–262.

Huber, M. T., Braun, H. A., and Krieg, J. C. (2000). Effects of noise on different disease states of recurrent affective disorders. *Biological Psychiatry*, 47(7):634–642.

Huesgen, C. T., Burger, P. C., Crain, B. J., and Johnson, G. A. (1993). In vitro mr microscopy of the hippocampus in Alzheimer's disease. *Neurology*, 43:145–152.

Hung, S. L. and Adeli, H. (1993). Parallel backpropagation learning algorithms on Cray Y-MP8/864 supercomputer. *Neurocomputing*, 5(6):287–302.

Iasemidis, L., Konstantinos, E., Principe, J. C., and Sackellares, J. (1995). Spatiotemporal dynamics of human epileptic seizures. In *Proceedings of the 3rd Experimental Chaos Conference*, pages 26–30, Edinburgh, Scotland, UK.

Iasemidis, L. D. (2003). Epileptic seizure prediction and control. *IEEE Transactions on Biomedical Engineering*, 50(5):549–556.

Iasemidis, L. D., Olson, L. D., Sackellares, J. C., and Savit, R. S. (1994). Time dependencies in the occurrences of epileptic seizures: A nonlinear approach. *Epilepsy Research*, 17(1):81–94.

Iasemidis, L. D., Principe, J. C., and Sackellares, J. C. (2000a). Measurement and quantification of spatiotemporal dynamics of human epileptic seizures. In Akay, M., editor, *Nonlinear Biomedical Signal Processing*, number II, pages 294–318. Wiley-IEEE Press, New York.

Iasemidis, L. D. and Sackellares, J. C. (1991). The temporal evolution of the largest Lyapunov exponent on the human epileptic cortex. In Duke, D. W. and Pritchard, W. S., editors, *Measuring Chaos in the Human Brain*, pages 49–82. World Scientific, Singapore.

Iasemidis, L. D., Shiau, D. S., Chaovalitwongse, W., Sackellares, J. C., Pardalos, P. M., Principe, J. C., Carney, P. R., Prasad, A., Veeramani, B., and Tsakalis, K. (2003). Adaptive epileptic seizure prediction system. *IEEE Transactions on Biomedical Engineering*, 50(5):616–627.

Iasemidis, L. D., Shiau, D. S., Sackellares, J. C., and Pardalos, P. M. (2000b). Transition to epileptic seizures: Optimization. *DIMACS Series in Discrete Mathematics and Theoretical Computer Science, American Mathematical Society*, 55:55–74.

Iglesias, J. and Villa, A. E. P. (2008). Emergence of preferred firing sequences in large spiking neural networks during simulated neuronal development. *International Journal of Neural Systems*, 18(4):267–277.

Ikawa, M., Nakanishi, M., Furukawa, T., Nakaaki, S., Hori, S., and Yoshida, S. (2000). Relationship between EEG dimensional complexity and neuropsychological findings in Alzheimer's disease. *Psychiatry and Clinical Neurosciences*, 54:537–541.

Isotani, T., Lehmann, D., Pascual-Marqui, R. D., Kochi, K., Wackermann, J., Saito, N., Yagyu, T., Kinoshita, T., and Sasada, K. (2001). EEG source localization and global dimensional complexity in high- and low-hypnotizable subjects: A pilot study. *Neuropsychobiology*, 44:192–198.

Izhikevich, E. M. (2001). Resonate-and-fire neurons. *Neural Networks*, 14:883–894.

Izhikevich, E. M. (2003). Simple model of spiking neurons. *IEEE Transactions on Neural Networks*, 14(6):1569–1572.

Izhikevich, E. M. (2007). Solving the distal reward problem through linkage of STDP and dopamine signaling. *Cerebral Cortex*, 17:2443–2452.

Izhikevich, E. M., Gally, J. A., and Edelman, G. M. (2004). Spike-timing dynamics of neuronal groups. *Cerebral Cortex*, 14(8):933–944.

Jack, Jr., C. R., Bentley, M. D., Twomey, C. K., and Zinsmeister, A. R. (1990). MR imaging-based volume measurements of the hippocampal formation and anterior temporal lobe: Validation studies. *Radiology*, 176:205–209.

Jaeger, H. (2001). The "echo state" approach to analysis and training recurrent neural networks, technical report tr-gmd-148. Technical report, German National Research Center for Information Technology.

Jaeger, H. and Lukosevicius, M. (2007). Optimization and applications of echo state networks with leaky-integrator neurons. *Neural Networks*, 20(3):335–352.

Jameson, L., Hussaini, M., and Earlbacher, M. (1996). *Wavelets Theory and Applications, ICASE/LaRC Series in Computational Science and Engineering*. Oxford University Press, NY.

Janke, A. L., de Zubicaray, G., Rose, S. E., Griffin, M., Chalk, J. B., and Galloway, G. J. (2001). 4D deformation modeling of cortical disease progression in Alzheimer's dementia. *Magnetic Resonance in Medicine*, 46:661–666.

Jelles, B., van Birgelen, J. H., Slaets, J. P. J., Hekster, R. E. M., Jonkman, E. J., and Stam, C. J. (1999). Decrease of non-linear structure in the EEG of Alzheimer's patients compared to healthy controls. *Clinical Neurophysiology*, 110:1159–1167.

Jeong, J. (2002). Nonlinear dynamics of EEG in Alzheimer's disease. *Drug Development Research*, 56:57–66.

Jeong, J. (2004). EEG dynamics in patients with Alzheimer's disease. *Clinical Neurophysiology*, 115:1490–1505.

Jeong, J., Chae, J. H., Kim, S. Y., and Han, S.-H. (2001a). Nonlinear dynamical analysis of the EEG in patients with Alzheimer's disease and vascular dementia. *Journal of Clinical Neurophysiology*, 18:58–67.

Jeong, J., Gore, J. C., and Peterson, B. S. (2001b). Mutual information analysis of the EEG in patients with Alzheimer's disease. *Clinical Neurophysiology*, 112:827–835.

Jeong, J., Kim, D. J., Kim, S. Y., Chae, J. H., Go, H. J., and Kim, K. S. (2001c). Effect of total sleep deprivation on the dimensional complexity of the waking EEG. *Sleep: Journal of Sleep and Sleep Disorders Research*, 24:197–202.

Jeong, J., Kim, M. S., and Kim, S. Y. (1999). Test for low-dimensional determinism in electroencephalograms. *Physical Review E*, 60:831–837.

Jeong, J., Kim, S. Y., and Han, S.-H. (1998). Nonlinear dynamic analysis of the EEG in Alzheimer's disease with optimal embedding dimension. *Electroencephalography and Clinical Neurophysiology*, 106:220–228.

Jiang, X. and Adeli, H. (2003). Fuzzy clustering approach for accurate embedding dimension identification in chaotic time series. *Integrated Computer-Aided Engineering*, 10(3):287–302.

Jiang, X. and Adeli, H. (2004). Wavelet packet-autocorrelation function method for traffic flow pattern analysis. *Computer-Aided Civil and Infrastructure Engineering*, 19(5):324–337.

Jiang, X. and Adeli, H. (2005a). Dynamic wavelet neural network for nonlinear identification of highrise buildings. *Computer-Aided Civil and Infrastructure Engineering*, 20(5):316–330.

Jiang, X. and Adeli, H. (2005b). Dynamic wavelet neural network model for traffic flow forecasting. *Journal of Transportation Engineering*, 131(10):771–779.

Jiang, X. and Adeli, H. (2008a). Dynamic fuzzy wavelet neuroemulator for nonlinear control of irregular highrise building structures. *International Journal for Numerical Methods in Engineering*, 74(7):1045–1066.

Jiang, X. and Adeli, H. (2008b). Neuro-genetic algorithm for nonlinear active control of highrise buildings. *International Journal for Numerical Methods in Engineering*, 75(8):770–786.

Jin, Y., Wen, R., and Sendhoff, B. (2007). Evolutionary multi-objective optimization of spiking neural networks. In Marques de Sa, J. P., Alexandre, L. A., Duch, W., and Mandic, D., editors, *Proceedings of the 17th International Conference on Artificial Neural Networks*, pages 370–379, Porto, Portugal. Springer.

Kachigan, S. K. (1984). *Statistical Analysis: An Interdisciplinary Introduction to Univariate and Multivariate Methods*. Radius Press, New York.

Kaiser, F. and Feldbusch, F. (2007). Building a bridge between spiking and artificial neural networks. In Marques de Sa, J. P., Alexandre, L. A., Duch, W., and Mandic, D., editors, *Proceedings of the 17th International Conference on Artificial Neural Networks*, pages 380–389, Porto, Portugal. Springer.

Kandel, E. R., Schwartz, J. H., and Jessell, T. M. (2000). *Principles of Neural Science*. McGraw-Hill, NY, 4th edition.

Karas, G. B., Burton, E. J., Rombouts, S. A., van Schijndel, R. A., O'Brien, J. T., Scheltens, P., McKeith, I. G., Williams, D., Ballard, C., and Barkhof, F. (2003). A comprehensive study of gray matter loss in patients with Alzheimer's disease using optimized voxel-based morphometry. *NeuroImage*, 18:895–907.

Karim, A. and Adeli, H. (2002a). Comparison of the fuzzy-wavelet RBFNN freeway incident detection model with the California algorithm. *Journal of Transportation Engineering*, 128(1):21–30.

Karim, A. and Adeli, H. (2002b). Incident detection algorithm using wavelet energy representation of traffic patterns. *Journal of Transportation Engineering*, 128(3):232–242.

Karim, A. and Adeli, H. (2003). Radial basis function neural network for work zone capacity and queue estimation. *Journal of Transportation Engineering*, 129(5):494–503.

Kasinski, A. and Ponulak, F. (2005). Experimental demonstration of learning properties of a new supervised learning method for the spiking neural networks. In *Proceedings of the 15th International Conference on Artificial Neural Networks: Biological Inspirations - Lecture Notes in Computer Science*, volume 3696, pages 145–153. Springer, Berlin.

Kellaway, P. (1990). An orderly approach to visual analysis: characteristics of the normal EEG of adults and children. In Daly, D. and Pedley, T., editors, *Current Practice of Clinical Electroencephalography*, pages 139–199. Raven Press, New York, NY, 2nd edition.

Kepler, T. B., Abbott, L. F., and Marder, E. (1992). Reduction of conductance-based neuron models. *Biological Cybernetics*, 66:381–387.

Khachaturian, Z. S. and Radebaugh, T. S. (1996). Synthesis of critical topics in Alzheimer's disease. In Khachaturian, Z. S. and Radebaugh, T. S., editors, *Alzheimer's Disease: Cause(s), Diagnosis, Treatment and Care*, pages 4–12. CRC Press, Boca Raton, FL.

Kim, H., Kim, S., Go, H., and Kim, D. (2001). Synergetic analysis of spatiotemporal EEG patterns: Alzheimer's disease. *Biological Cybernetics*, 85:1–17.

Kistler, W. M., Gerstner, W., and van Hemmen, J. L. (1997). Reduction of Hodgkin-Huxley equations to a single-variable threshold model. *Neural Computation*, 9:1015–1045.

Kobayashi, T., Madokoro, S., Wada, Y., Misaki, K., and Nakagawa, H. (2001). Human sleep EEG analysis using correlation dimension analysis. *Clinical Electroencephalography*, 32(3):112–118.

Kobayashi, T., Madokoro, S., Wada, Y., Misaki, K., and Nakagawa, H. (2002). Effect of ethanol on human sleep EEG using correlation dimension analysis. *Neuropsychobiology*, 46(2):104–110.

Kohonen, T. (1982). Self-organized formation of topologically correct feature maps. *Biological Cybernetics*, 43:59–69.

Kondakor, I., Michel, C. M., Wackermann, J., Koenig, T., Tanaka, H., Peuvot, J., and Lehmann, D. (1999). Single-dose piracetam effects on global complexity measures of human spontaneous multichannel EEG. *International Journal of Psychophysiology*, 34:81–87.

Kotagal, P. (2001). Neocortical temporal lobe epilepsy. In Lders, H. O. and Comair, Y. G., editors, *Epilepsy Surgery*, pages 105–109. Lippincott Williams & Wilkins, Philadelphia, PA.

Krishnan, K. R., Charles, H. C., Doraiswamy, P. M., Mintzer, J., Weisler, R., Yu, X., Perdomo, C., Ieni, J. R., and Rogers, S. (2003). Randomized, placebo-controlled trial of the effects of donepezil on neuronal markers and hippocampal volumes in Alzheimer's disease. *American Journal of Psychiatry*, 160:2003–2011.

Lee, H. and Choi, S. (2003). PCA+HMM+SVM for EEG pattern classification. In *Proceedings of Seventh International Symposium on Signal Processing and Its Applications*, volume 1, pages 541–544.

Lee, Y. J., Zhu, Y. S., Xu, Y. H., Shen, M. F., Tong, S. B., and Thakor, N. V. (2001). The nonlinear dynamical analysis of the EEG in schizophrenia with temporal and spatial embedding dimension. *Journal of Medical Engineering and Technology*, 25:79–83.

Legenstein, R., Naeger, C., and Maass, W. (2005). What can a neuron learn with spike-timing-dependent plasticity? *Neural Computation*, 17(11):2337–2382.

Lehmann, D. and Michel, C. M. (1990). Intracerebral dipole source localization for FFT power maps. *Electroencephalography and Clinical Neurophysiology*, 76:271–276.

Leuchter, A. F., Cook, I. A., Lufkin, R. B., Dunkin, J., Newton, T. F., Cummings, J. L., Mackey, J. K., and Walter, D. O. (1994). Cordance: A new

. method for assessment of cerebral perfusion and metabolism using quantitative electroencephalography. *NeuroImage*, 1:208–219.

Leuchter, A. F., Cook, I. A., Newton, T. F., Dunkin, J., Walter, D. O., Rosenberg-Thompson, S., Lachenbruch, P. A., and Weiner, H. (1993). Regional differences in brain electrical activity in dementia: Use of spectral power and spectral ratio measures. *Electroencephalography and Clinical Neurophysiology*, 87:385–393.

Liew, A. W. and Yan, H. (2003). An adaptive spatial fuzzy clustering algorithm for 3-D MR image segmentation. *IEEE Transactions on Medical Imaging*, 22:1063–1075.

Litt, B. and Echauz, J. (2002). Prediction of epileptic seizures. *The Lancet Neurology*, 1(1):22–30.

Liu, H., Wang, X., and Qiang, W. (2007). A fast method for implicit surface reconstruction based on radial basis functions network from 3D scattered points. *International Journal of Neural Systems*, 17(6):459–465.

Locatelli, T., Cursi, M., Liberati, D., Franceschi, M., and Comi, G. (1998). EEG coherence in Alzheimer's disease. *Electroencephalography and Clinical Neurophysiology*, 106:229–237.

Lopes da Silva, F. H., Pijn, J. P., and Wadman, W. J. (1994). Dynamics of local neuronal networks: Control parameters and state bifurcations in epileptogenesis. *Progress in Brain Research*, 102:359–370.

Maass, W. (1996). Lower bounds for the computational power of spiking neural networks. *Neural Computation*, 8(1):1–40.

Maass, W. (1997a). Fast sigmoidal networks via spiking neurons. *Neural Computation*, 9(2):279–304.

Maass, W. (1997b). Networks of spiking neurons: The third generation of spiking neural network models. *Neural Networks*, 10(9):1659–1671.

Maass, W. (1997c). Noisy spiking neurons with temporal coding have more computational power than sigmoidal neurons. In Mozer, M., Jordan, M. I., and Petsche, T., editors, *Advances in Neural Information Processing Systems, Vol. 9*, pages 211–217. MIT Press, Cambridge, MA.

Maass, W. and Natschläger, T. (1997). Networks of spiking neurons can emulate arbitrary Hopfield nets in temporal coding. *Network: Computation in Neural Systems*, 8(4):355–372.

Maass, W. and Natschläger, T. (1998a). Associative memory with networks of spiking neurons in temporal coding. In Smith, L. S. and Hamilton, A., editors, *Neuromorphic Systems: Engineering Silicon from Neurobiology*, pages 21–32. World Scientific.

Maass, W. and Natschläger, T. (1998b). Emulation of Hopfield networks with spiking neurons in temporal coding. In Bower, J. M., editor, *Computational Neuroscience: Trends in Research*, pages 221–226. Plenum Press.

Maass, W., Natschläger, T., and Markram, H. (2002). Real-time computing without stable states: A new framework for neural computation based on perturbations. *Neural Computation*, 14(11):2531–2560.

Mallat, S. G. (1989). A theory for multi-resolution signal decomposition: The wavelet representation. *IEEE Transactions on Pattern Analysis and Machine Intelligence*, 11(7):674–693.

Mallat, S. G. (1998). *Wavelet Tour of Signal Processing*. Academic Press, London.

Matousek, M., Wackermann, J., Palus, M., Berankova, A., Albrecht, V., and Dvorak, I. (1995). Global dimensional complexity of the EEG in healthy volunteers. *Neuropsychobiology*, 31:47–52.

Mattia, D., Babiloni, F., Romigi, A., Cincotti, F., Bianchi, L., Sperli, F., Placidi, F., Bozzao, A., Giacomini, P., Floris, R., and Grazia Marciani, M. (2003). Quantitative EEG and dynamic susceptibility contrast MRI in Alzheimer's disease: A correlative study. *Clinical Neurophysiology*, 114:1210–1216.

Mayeux, R. (1996). Putative risk factors for Alzheimer's disease. In Khachaturian, Z. S. and Radebaugh, T. S., editors, *Alzheimer's Disease: Cause(s), Diagnosis, Treatment and Care*, pages 40–49. CRC Press, Boca Raton, FL.

Mayorga, R. and Carrera, J. (2007). A radial basis function network approach for the computation of inverse continuous time variant functions. *International Journal of Neural Systems*, 17(3):149–160.

Mazzoni, P., Andersen, R. A., and Jordan, M. I. (1991). More biologically plausible learning rule than backpropagation applied to a network model of cortical area 7a. *Cerebral Cortex*, 1:293–307.

McKennoch, S., Liu, D., and Bushnell, L. G. (2006). Fast modifications of the SpikeProp algorithm. In *Proceedings of the International Joint Conference on Neural Networks*, pages 3970–3977, Vancouver, Canada.

McKennoch, S., Voegtlin, T., and Bushnell, L. (2009). Spike-timing error backpropagation in theta neuron networks. *Neural Computation*, 21(1):9–45.

Menschik, E. D. and Finkel, L. H. (1998). Neuromodulatory control of hippocampal function: Towards a model of Alzheimer's disease. *Artificial Intelligence in Medicine*, 13:99–121.

Menschik, E. D. and Finkel, L. H. (1999). Cholinergic neuromodulation and Alzheimer's disease: From single cells to network simulations. *Progress in Brain Research*, 121:19–45.

Menschik, E. D. and Finkel, L. H. (2000). Cholinergic neuromodulation of an anatomically reconstructed hippocampal ca3 pyramidal cell. *Neurocomputing*, 32-33:197–205.

Menschik, E. D., Yen, S.-C., and Finkel, L. H. (1999). Model- and scale-independent performance of a hippocampal ca3 network architecture. *Neurocomputing*, 26-27:443–453.

Mesulam, M. (2004). The cholinergic lesion of Alzheimer's disease: Pivotal factor or side show? *Learning and Memory (Cold Spring Harbor, NY)*, 11:43–49.

Meyer, Y. (1993). *Wavelets: Algorithms and Applications*. Society for Industrial and Applied Mathematics, Philadelphia, PA.

Meyer-Lindenberg, A. (1996). The evolution of complexity in human brain development: An EEG study. *Electroencephalography and Clinical Neurophysiology*, 99:405–411.

Meyer-Lindenberg, A., Bauer, U., Krieger, S., Lis, S., Vehmeyer, K., Schuler, G., and Gallhofer, B. (1998). The topography of non-linear cortical dynamics at rest, in mental calculation and moving shape perception. *Brain Topography*, 10:291–299.

Michel, C. M., Henggeler, B., Brandeis, D., and Lehmann, D. (1993). Localization of sources of brain alpha/theta/delta activity and the influence of the mode of spontaneous mentation. *Physiological Measurement*, 14:A21–A26.

Migliore, M., Cook, E. P., Jaffe, D. B., Turner, D. A., and Johnston, D. (1995). Computer simulations of morphologically reconstructed CA3 hippocampal neurons. *Journal of Neurophysiology*, 73:1157–1168.

Mirra, S. S. and Markesbery, W. R. (1996). The neuropathology of Alzheimer's disease: Diagnostic features and standardization. In Khachaturian, Z. S. and Radebaugh, T. S., editors, *Alzheimer's Disease: Cause(s), Diagnosis, Treatment and Care*, pages 111–123. CRC Press, Boca Raton, FL.

Miyauchi, T., Hagimoto, H., Ishii, M., Endo, S., Tanaka, K., Kajiwara, S., Endo, K., Kajiwara, A., and Kosaka, K. (1994). Quantitative EEG in patients with presenile and senile dementia of the Alzheimer type. *Acta Neurologica Scandinavica*, 89:56–64.

Mohamed, N., Rubin, D. M., and Marwala, T. (2006). Detection of epileptiform activity in human EEG signals using Bayesian neural networks. *Neural Information Processing - Letters and Reviews*, 10(1):1–10.

Molle, M., Marshall, L., Wolf, B., Fehm, H. L., and Born, J. (1999). EEG complexity and performance measures of creative thinking. *Psychophysiology*, 36:95–104.

Molnar, M. and Skinner, J. E. (1991). Correlation dimension changes of the EEG during the wakefulness-sleep cycle. *Acta Biochimica et Biophysica Hungarica*, 26(1-4):121–125.

Montplaisir, J., Petit, D., Gauthier, S., Gaudreau, H., and Dcary, A. (1998). Sleep disturbances and EEG slowing in Alzheimer's disease. *Sleep Research Online*, 1:147–151.

Moore, S. C. (2002). Backpropagation in spiking neural networks. Master's thesis, University of Bath.

Morris, J. C., Storandt, M., Miller, J. P., McKeel, D. W., Price, J. L., Rubin, E. H., and Berg, L. (2001). Mild cognitive impairment represents early-stage Alzheimer disease. *Archives of Neurology*, 58:397–405.

Musha, T., Asada, T., Yamashita, F., Kinoshita, T., Chen, Z., Matsuda, H., Uno, M., and Shankle, W. R. (2002). A new EEG method for estimating cortical neuronal impairment that is sensitive to early stage Alzheimer's disease. *Clinical Neurophysiology*, 113:1052–1058.

Najm, I. M., Babb, T. L., Mohamed, A., Diehl, B., Ng, T. C., Bingaman, W. E., and Lders, H. O. (2001). Mesial temporal lobe sclerosis. In Lders, H. O. and Comair, Y. G., editors, *Epilepsy Surgery*, pages 95–103. Lippincott Williams & Wilkins, Philadelphia, PA.

Natarajan, K., Acharya, R. U., Alias, F., Tiboleng, T., and Puthusserypady, S. K. (2004). Nonlinear analysis of EEG signals at different mental states. *BioMedical Engineering OnLine*, 3(7):1–11.

National Institute of Neurological Disorders and Stroke (2004). Seizures and epilepsy: Hope through research. NINDS, Bethesda, MD. Available: `http://www.ninds.nih.gov/disorders/epilepsy`.

Natschläger, T. and Ruf, B. (1998). Spatial and temporal pattern analysis via spiking neurons. *Network: Computation in Neural Systems*, 9(3):319–332.

Natschläger, T. and Ruf, B. (1999). Pattern analysis with spiking neurons using delay coding. *Neurocomputing*, 26-27(1-3):463–469.

Natschläger, T., Ruf, B., and Schmitt, M. (2001). Unsupervised learning and self-organization in networks of spiking neurons. In Seiffert, U. and Jain, L. C., editors, *Self-Organizing Neural Networks. Recent Advances and Applications*, volume 78 of *Springer Series on Studies in Fuzziness and Soft Computing*. Springer-Verlag, Heidelberg.

Newland, D. (1993). *An Introduction to Random Vibrations, Spectral & Wavelet Analysis*. John Wiley & Sons, New York, NY.

Newman, D. J., Hettich, S., Blake, C. L., and Merz, C. J. (1998). UCI repository of machine learning databases. Department of Information and Computer Science, University of California, Irvine, CA, http://www.ics.uci.edu/mlearn/MLRepository.html.

Niedermeyer, E. (1999). Historical aspects. In Niedermeyer, E. and Silva, F., editors, *Electroencephalography, Basic Principles, Clinical Applications, and Related Fields*, pages 1–14. Williams & Wilkins, Baltimore, MD.

Niestroj, E., Spieweg, I., and Herrmann, W. M. (1995). On the dimensionality of sleep-EEG data: Using chaos mathematics and a systematic variation of the parameters of the corex program to determine the correlation exponents of sleep EEG segments. *Neuropsychobiology*, 31(3):166–172.

Notley, S. V. and Elliott, S. J. (2003). Efficient estimation of a time-varying dimension parameter and its application to EEG analysis. *IEEE Transactions on Biomedical Engineering*, 50(5):594–602.

Osborne, A. R. and Provenzale, A. (1989). Finite correlation dimension for stochastic systems with power-law spectra. *Physica D*, 35:357–381.

Palus, M. (1996). Nonlinearity in normal human EEG: Cycles, temporal asymmetry, nonstationarity and randomness, not chaos. *Biological Cybernetics*, 75:389–396.

Panakkat, A. and Adeli, H. (2007). Neural network models for earthquake magnitude prediction using multiple seismicity indicators. *International Journal of Neural Systems*, 17(1):13–33.

Panuku, L. N. and Sekhar, C. C. (2007). Clustering of nonlinearly separable data using spiking neural networks. In Marques de Sa, J. P., Alexandre, L. A., Duch, W., and Mandic, D., editors, *Proceedings of the 17th International Conference on Artificial Neural Networks*, pages 390–399, Porto, Portugal. Springer.

Park, H. S. and Adeli, H. (1997). Distributed neural dynamics algorithms for optimization of large steel structures. *Journal of Structural Engineering*, 123(7):880–888.

Paugam-Moisy, H. and Bohte, S. (2009). Computing with spiking neuron networks. In Kok, J. and Heskes, T., editors, *Handbook of Natural Computing*. Springer-Verlag.

Paulus, M. P. and Braff, D. L. (2003). Chaos and schizophrenia: Does the method fit the madness? *Biological Psychiatry*, 53(1):3–11.

Paulus, M. P., Geyer, M. A., and Braff, D. L. (1996). Use of methods from chaos theory to quantify a fundamental dysfunction in the behavioral organization of schizophrenic patients. *American Journal of Psychiatry*, 153(5):714–717.

Pavlidis, N. G., Tasoulis, D. K., Plagianakos, V. P., and Vrahatis, M. N. (2005). Spiking neural network training using evolutionary algorithms. In *Proceedings of the International Joint Conference on Neural Networks*, pages 2190–2194, Montreal, QC.

Pearlson, G. D., Harris, G. J., Powers, R. E., Barta, P. E., Camargo, E. E., Chase, G. A., Noga, J. T., and Tune, L. E. (1992). Quantitative changes in mesial temporal volume, regional cerebral blood flow, and cognition in Alzheimer's disease. *Archives of General Psychiatry*, 49:402–408.

Pedrycz, W., Rai, R., and Zurada, J. (2008). Experience-consistent modeling for radial basis function neural networks. *International Journal of Neural Systems*, 18(4):279–292.

Pereda, E., Gamundi, A., Rial, R., and Gonzalez, J. (1998). Non-linear behavior of human EEG: Fractal exponent versus correlation dimension in awake and sleep stages. *Neuroscience Letters*, 250:91–94.

Perez de Alejo, R., Ruiz-Cabello, J., Cortijo, M., Rodriguez, I., Echave, I., Regadera, J., Arrazola, J., Aviles, P., Barreiro, P., Gargallo, D., and Grana, M. (2003). Computer-assisted enhanced volumetric segmentation magnetic resonance imaging data using a mixture of artificial neural networks. *Magnetic Resonance Imaging*, 21:901–912.

Petrella, J. R., Coleman, R. E., and Doraiswamy, P. M. (2003). Neuroimaging and early diagnosis of Alzheimer's disease: A look to the future. *Radiology*, 226:315–336.

Petrosian, A., Homan, R., Prokhorov, D., and Wunsch, D. (1996). Classification of epileptic EEG using neural network and wavelet transform. In Unser, M. A., Aldroubi, A., and Laine, A. F., editors, *SPIE Proceedings of Conference on Wavelet Applications in Signal and Image Processing IV*, pages 834–843, Bellingham, WA. SPIE.

Petrosian, A., Prokhorov, D., Homan, R., Dascheiff, R., and Wunsch, D. (2000a). Recurrent neural network based prediction of epileptic seizures in intra- and extra-cranial EEG. *Neurocomputing*, 30(1-4):201–218.

Petrosian, A., Prokhorov, D., Lajara-Nanson, W., and Schiffer, R. (2001). Recurrent neural network-based approach for early recognition of Alzheimer's disease in EEG. *Clinical Neurophysiology*, 112(8):1378–1387.

Petrosian, A., Prokhorov, D., and Schiffer, R. (2000b). Early recognition of Alzheimer's disease in EEG using recurrent neural network and wavelet transform. In Aldroubi, A., Laine, A. F., and Unser, M. A., editors, *SPIE Proceedings of Conference on Wavelet Applications in Signal and Image Processing VIII, San Diego, CA*, pages 870–877, Bellingham, WA. SPIE.

Pezard, L., Lachaux, J. P., Thomasson, N., and Martinerie, J. (1999). Why bother to spatially embed EEG? Comments on Pritchard et al., Psychophysiology, 1996, 33, 362-368. *Psychophysiology*, 36:527–531.

Pijn, J. P. M., van Neerven, J., Noest, A., and Lopes da Silva, F. (1991). Chaos or noise in EEG signals: Dependence on state and brain site. *Electroencephalography and Clinical Neurophysiology*, 79:371–381.

Pinsky, P. F. and Rinzel, J. (1994). Intrinsic and network rhythmogenesis in a reduced Traub model for CA3 neurons. *Journal of Computational Neuroscience*, 1:39–60.

Pizzi, N., Choo, L. P., Mansfield, J., Jackson, M., Halliday, W. C., Mantsch, H. H., and Somorjai, R. L. (1995). Neural network classification of infrared spectra of control and Alzheimer's diseased tissue. *Artificial Intelligence in Medicine*, 7:67–79.

Planel, E., Miyasaka, T., Launey, T., Chui, D. H., Tanemura, K., Sato, S., Murayama, O., Ishiguro, K., Tatebayashi, Y., and Takashima, A. (2004). Alterations in glucose metabolism induce hypothermia leading to tau hyperphosphorylation through differential inhibition of kinase and phosphatase activities: Implications for Alzheimer's disease. *Journal of Neuroscience*, 24:2401–2411.

Polikar, R., Greer, M. H., Udpa, L., and Keinert, F. (1997). Multiresolution wavelet analysis of ERPs for the detection of Alzheimer's disease. In *Proceedings of the 19th International Conference of the IEEE Engineering in Medicine and Biology Society, Chicago, IL*, pages 1301–1304, New York, NY. IEEE.

Porter, R. J. (1993). Classification of epileptic seizures and epileptic syndromes. In Laidlaw, J., Richens, A., and Chadwick, D., editors, *A Textbook of Epilepsy*, pages 1–22. Churchill Livingstone, London, 4th edition.

Price, D. L. (2000). Aging of the brain and dementia of the Alzheimer type. In Kandel, E., Schwartz, J., and Jessel, T., editors, *Principles of Neural Science*, pages 1149–1161. 4th edition, New York.

Prichep, L. S., John, E. R., Ferris, S. H., Reisberg, B., Almas, M., Alper, K., and Cancro, R. (1994). Quantitative EEG correlates of cognitive deterioration in the elderly. *Neurobiology of Aging*, 15:85–90.

Pritchard, W. S. (1999). On the validity of spatial embedding: A reply to Pezard et al., Psychophysiology, 1999, 36, 527-531. *Psychophysiology*, 36:532–535.

Pritchard, W. S., Duke, D. W., and Coburn, K. L. (1991). Altered EEG dynamical responsivity associated with normal aging and probable Alzheimer's disease. *Dementia*, 2:102–105.

Pritchard, W. S., Duke, D. W., Coburn, K. L., Moore, N. C., Tucker, K. A., Jann, M. W., and Hostetler, R. M. (1994). EEG-based, neural-net predictive classification of Alzheimer's disease versus control subjects is augmented by non-linear EEG measures. *Electroencephalography and Clinical Neurophysiology*, 91(2):118–130.

Pritchard, W. S., Duke, D. W., and Krieble, K. K. (1995). Dimensional analysis of resting human EEG. II: Surrogate-data testing indicates nonlinearity but not low-dimensional chaos. *Psychophysiology*, 32:486–491.

Pritchard, W. S., Krieble, K. K., and Duke, D. W. (1996). On the validity of estimating EEG correlation dimension from a spatial embedding. *Psychophysiology*, 33:362–368.

Pucci, E., Belardinelli, N., Cacchi, G., Signorino, M., and Angeleri, F. (1999). EEG power spectrum differences in early and late onset forms of Alzheimer's disease. *Clinical Neurophysiology*, 110:621–631.

Radau, P. E., Slomka, P. J., Julin, P., Svensson, L., and Wahlund, L. O. (2001). Evaluation of linear registration algorithms for brain SPECT and the errors due to hypoperfusion lesions. *Medical Physics*, 28:1660–1668.

Rao, R. M. and Bopardikar, A. S. (1998). *Wavelet Transforms: Introduction to Theory and Applications*. Addison-Wesley, Reading, MA.

Rapp, P. E., Albano, A. M., Schmah, T. I., and Farwell, L. A. (1993). Filtered noise can mimic low-dimensional chaotic attractors. *Physical Review E*, 47:2289–2297.

Reddick, W. E., Glass, J. O., Cook, E. N., Elkin, T. D., and Deaton, R. J. (1997). Automated segmentation and classification of multispectral magnetic resonance images of brain using artificial neural networks. *IEEE Transactions on Medical Imaging*, 16:911–918.

Reed, B. R., Jagust, W. J., Sacb, J. P., and Ober, B. A. (1989). Memory and regional cerebral blood flow in mildly symptomatic Alzheimer's disease. *Neurology*, 39:1537–1539.

Reggia, J. A., Ruppin, E., and Berndt, R. S. (1997). Computer models: A new approach to the investigation of disease. *M.D. Computing: Computers in Medical Practice*, 14:160–168.

Riedmiller, M. and Braun, H. (1993). A direct adaptive method for faster backpropagation learning: The Rprop algorithm. In *IEEE International Conference on Neural Networks*, volume 1, pages 586–591, San Francisco, CA.

Rigatos, G. G. (2008). Adaptive fuzzy control with output feedback for H-infinity tracking of SISI nonlinear systems. *International Journal of Neural Systems*, 18(4):305–320.

Rinzel, J. and Ermentrout, G. B. (1989). Analysis of neuronal excitability and oscillations. In Koch, C. and Segev, I., editors, *Methods in Neuronal Modeling*, pages 135–169. MIT Press, Cambridge, MA.

Rodin, E. and Ancheta, O. (1987). Cerebral electrical fields during petit mal absences. *Electroencephalography and Clinical Neurophysiology*, 66:457–466.

Rombouts, S. A., Barkhof, F., Witter, M. P., and Scheltens, P. (2000). Unbiased whole-brain analysis of gray matter loss in Alzheimer's disease. *Neuroscience Letters*, 285:231–233.

Rombouts, S. A. R. B., Keunen, R. W. M., and Stam, C. J. (1995). Investigation of nonlinear structure in multichannel EEG. *Physics Letters A*, 202:352–358.

Roschke, J. and Aldenhoff, J. B. (1991). The dimensionality of human's electroencephalogram during sleep. *Biological Cybernetics*, 64(4):307–313.

Roschke, J. and Aldenhoff, J. B. (1993). Estimation of the dimensionality of sleep-EEG data in schizophrenics. *European Archives of Psychiatry and Clinical Neuroscience*, 242(4):191–196.

Rosenstein, M. T., Collins, J. J., and De Luca, C. J. (1993). A practical method for calculating largest Lyapunov exponents from small data sets. *Physica D*, 65(1-2):117–134.

Rumelhart, D. E., Hinton, G. E., and Williams, R. J. (1986). Learning internal representations by error propagation. In Rumelhart, D. E. and McClelland, J. L., editors, *Parallel Distributed Processing, Vol. 1*, pages 318–362. MIT Press, Cambridge, MA.

Ruppin, E. and Reggia, J. A. (1995). A neural model of memory impairment in diffuse cerebral atrophy. *British Journal of Psychiatry*, 166:19–28.

Sabourin, C., Madani, K., and Bruneau, O. (2007). Autonomous biped gait pattern based on fuzzy-CMAC neural networks. *Integrated Computer-Aided Engineering*, 14(2):173–186.

Saito, N., Kuginuki, T., Yagyu, T., Kinoshita, T., Koenig, T., Pascual-Marqui, R. D., Kochi, K., Wackermann, J., and Lehmann, D. (1998). Global, regional, and local measures of complexity of multichannel electroencephalography in acute, neuroleptic-naive, first-break schizophrenics. *Biological Psychiatry*, 43:794–802.

Samant, A. and Adeli, H. (2000). Feature extraction for traffic incident detection using wavelet transform and linear discriminant analysis. *Computer-Aided Civil and Infrastructure Engineering*, 15(4):241–250.

Samant, A. and Adeli, H. (2001). Enhancing neural network incident detection algorithms using wavelets. *Computer-Aided Civil and Infrastructure Engineering*, 16(4):239–245.

Sandson, T. A., Felician, O., Edelman, R. R., and Warach, S. (1999). Diffusion-weighted magnetic resonance imaging in Alzheimer's disease. *Dementia and Geriatric Cognitive Disorders*, 10:166–171.

Saykin, A. J. and Wishart, H. A. (2003). Mild cognitive impairment: Conceptual issues and structural and functional brain correlates. *Seminars in Clinical Neuropsychiatry*, 8:12–30.

Schaefer, A. M. and Zimmermann, H. G. (2007). Recurrent neural networks are universal approximators. *International Journal of Neural Systems*, 17(4):253–263.

Schrauwen, B. and van Campenhout, J. (2004). Extending SpikeProp. In *Proceedings of the International Joint Conference on Neural Networks*, pages 471–476, Budapest.

Schrauwen, B., Verstraeten, D., and van Campenhout, J. (2007). An overview of reservoir computing: theory, applications and implementations. In *Proceedings of the 15th European Symposium on Artificial Neural Networks - Advances in Computational Intelligence and Learning*, pages 471–482.

Sejnowski, T. J. (1986). Open questions about computation in the cerebral cortex. In Rumelhart, D. E. and McClelland, J. L., editors, *Parallel Distributed Processing*, pages 372–389. MIT Press, Cambridge, MA.

Selkoe, D. J. (1994). Alzheimer's disease: A central role for amyloid. *Journal of Neuropathology and Experimental Neurology*, 53:438–447.

Shah, Y., Tangalos, E. G., and Petersen, R. C. (2000). Mild cognitive impairment: When is it a precursor to Alzheimer's disease? *Geriatrics*, 55:62–68.

Shen, Y., Olbrich, E., Achermann, P., and Meier, P. F. (2003). Dimensional complexity and spectral properties of the human sleep EEG electroencephalograms. *Clinical Neurophysiology*, 114(2):199–209.

Siegle, G. J. and Hasselmo, M. E. (2002). Using connectionist models to guide assessment of psychological disorder. *Psychological Assessment*, 14:263–278.

Signorino, M., Brizioli, E., Amadio, L., Belardinelli, N., Pucci, E., and Angeleri, F. (1996). An EEG power index (eyes open vs. eyes closed) to differentiate Alzheimer's from vascular dementia and healthy ageing. *Archives of Gerontology and Geriatrics*, 22:245–260.

Signorino, M., Pucci, E., Belardinelli, N., Nolfe, G., and Angeleri, F. (1995). EEG spectral analysis in vascular and Alzheimer dementia. *Electroencephalography and Clinical Neurophysiology*, 94:313–325.

Silva, S. M. and Ruano, A. E. (2005). Application of Levenberg-Marquardt method to the training of spiking neural networks. In *Proceedings of the International Conference on Neural Networks and Brain*, volume 3, pages 1354–1358.

Simeoni, R. J. and Mills, P. M. (2003). Bispectral analysis of Alzheimer's electroencephalogram: A preliminary study. In Allen, B. and Lovel, N., editors, *International Federation for Medical and Biological Engineering (IFMBE) Proceedings of the World Congress on Medical Physics and Biomedical Engineering, Sydney, Australia*, Stockholm, Sweden. IFMBE.

Sirca, G. and Adeli, H. (2001). Neural network model for uplift load capacity of metal roof panels. *Journal of Structural Engineering*, 127(11):1276–1285.

Sirca, G. and Adeli, H. (2003). Neural network model for uplift load capacity of metal roof panels - closure. *Journal of Structural Engineering*, 129(4):562–563.

Slomka, P. J., Radau, P., Hurwitz, G. A., and Dey, D. (2001). Automated three-dimensional quantification of myocardial perfusion and brain SPECT. *Computerized Medical Imaging and Graphics*, 25:153–164.

Smith, L. I. (2002). A tutorial on principal components analysis. Available: http://csnet.otago.ac.nz/cosc453/student_tutorials/principal_components.pdf.

Smith, P. (1998). *Explaining Chaos*. Cambridge University Press, Cambridge, UK.

Soininen, H., Partanen, V. J., Laulumaa, V., Paakkonen, A., Helkala, E. L., and Riekkinen, P. J. (1991). Serial EEG in Alzheimer's disease: 3 year follow-up and clinical outcome. *Electroencephalography and Clinical Neurophysiology*, 79:342–348.

Soininen, H., Reinikainen, K. J., Partanen, V. J., Helkala, E. L., Paljarvi, L., and Riekkinen, P. J. (1992). Slowing of electroencephalogram and choline acetyltransferase activity in post mortem frontal cortex in definite Alzheimer's disease. *Neuroscience*, 49:529–535.

Sporns, O., Tononi, G., and Edelman, G. M. (2000). Connectivity and complexity: The relationship between neuroanatomy and brain dynamics. *Neural Networks*, 13:909–922.

Stam, C. J., Jelles, B., Achtereekte, H. A. M., Rombouts, S. A., Slaets, J. P. J., and Keunen, R. W. M. (1995). Investigation of EEG non-linearity in dementia and Parkinson's disease. *Electroencephalography and Clinical Neurophysiology*, 95:309–317.

Stam, C. J., Tavy, D. L. J., Jelles, B., Achtereekte, H. A. M., Slaets, J. P. J., and Keunen, R. W. M. (1994). Non-linear dynamical analysis of multi channel EEG data: Clinical applications in dementia and Parkinsons disease. *Brain Topography*, 7:141–150.

Stam, C. J., van Woerkom, T. C., and Pritchard, W. S. (1996). Use of nonlinear EEG measures to characterize EEG changes during mental activity. *Electroencephalography and Clinical Neurophysiology*, 99:214–224.

Stevens, A., Kircher, T., Nickola, M., Bartels, M., Rosellen, N., and Wormstall, H. (2001). Dynamic regulation of EEG power and coherence is lost early and globally in probable DAT. *European Archives of Psychiatry and Clinical Neuroscience*, 251:199–204.

Stork, D. G. (1989). Is backpropagation biologically plausible? In *Proceedings of the International Joint Conference on Neural Networks*, volume 2, pages 241–246, New York. IEEE Press.

Strang, G. (1996). *Wavelets and Filter Banks*. Wellesley-Cambridge Press, Wellesley, MA.

Sulimov, A. V. and Maragei, R. A. (2003). Sleep EEG as a nonlinear dynamic process: A comparison of global correlation dimension of human EEG and measures of linear interdependence between channels. *Zhurnal Vysshei Nervnoi Deiatelnosti Imeni I P Pavlova*, 53:151–155.

Szelenberger, W., Wackermann, J., Skalski, M., Drojewski, J., and Niemcewicz, S. (1996a). Interhemispheric differences of sleep EEG complexity. *Acta Neurobiologiae Experimentalis*, 56:955–959.

Szelenberger, W., Wackermann, J., Skalski, M., Niemcewicz, S., and Drojewski, J. (1996b). Analysis of complexity of EEG during sleep. *Acta Neurobiologiae Experimentalis*, 56:165–169.

Takens, F. (1981). Detecting strange attractors in turbulence: Dynamical systems and turbulence. In Rand, D. A. and Young, L. S., editors, *Lecture Notes in Mathematics, Vol. 366*. Springer-Verlag, Berlin.

Teipel, S. J., Bayer, W., Alexander, G. E., Bokde, A. L., Zebuhr, Y., Teichberg, D., Muller-Spahn, F., Schapiro, M. B., Moller, H. J., Rapoport, S. I., and Hampel, H. (2003). Regional pattern of hippocampus and corpus callosum atrophy in Alzheimer's disease in relation to dementia severity: Evidence for early neocortical degeneration. *Neurobiology of Aging*, 24:85–94.

Terry, R. D. (1996). A history of the morphology of Alzheimer's disease. In Khachaturian, Z. S. and Radebaugh, T. S., editors, *Alzheimer's Disease:*

Cause(s), Diagnosis, Treatment, and Care, pages 31–35. CRC Press, Boca Raton, FL.

Theiler, J., Eubank, S., Longtin, A., Galdrikian, B., and Farmer, J. D. (1992). Testing for nonlinearity in time series: The method of surrogate data. *Physica D*, 58:77–94.

Theiler, J. and Rapp, P. E. (1996). Re-examination of the evidence for low-dimensional, nonlinear structure in the human electroencephalogram. *Electroencephalography and Clinical Neurophysiology*, 98:213–222.

Thompson, P. M., Hayashi, K. M., de Zubicaray, G. I., Janke, A. L., Rose, S. E., Semple, J., Hong, M. S., Herman, D. H., Gravano, D., Doddrell, D. M., and Toga, A. W. (2004). Mapping hippocampal and ventricular change in Alzheimer disease. *NeuroImage*, 22:1754–1766.

Tian, J., Shi, J., Bailey, K., and Mann, D. M. (2003). Negative association between amyloid plaques and cerebral amyloid angiopathy in Alzheimer's disease. *Neuroscience Letters*, 352:137–140.

Tononi, G. and Edelman, G. M. (1998). Consciousness and complexity. *Science (Washington)*, 282:1846–1851.

Tononi, G., Edelman, G. M., and Sporns, O. (1998). Complexity and coherency: Integrating information in the brain. *Trends in Cognitive Sciences*, 2:474–484.

Tononi, G., Sporns, O., and Edelman, G. M. (1994). A measure for brain complexity: Relating functional segregation and integration in the nervous system. *Proceedings of the National Academy of Sciences*, 91:5033–5037.

Tononi, G., Sporns, O., and Edelman, G. M. (1996). A complexity measure for selective matching of signals by the brain. *Proceedings of the National Academy of Sciences*, 93:3422–3427.

Traub, R. D., Jefferys, J. G., Miles, R., Whittington, M. A., and Toth, K. (1994). A branching dendritic model of a rodent CA3 pyramidal neuron. *The Journal of Physiology*, 481:79–95.

Traub, R. D., Wong, R. K., Miles, R., and Michelson, H. (1991). A model of a CA3 hippocampal pyramidal neuron incorporating voltage-clamp data on intrinsic conductances. *Journal of Neurophysiology*, 66:635–650.

Turkheimer, F. E., Aston, J. A., Banati, R. B., Riddell, C., and Cunningham, V. J. (2003). A linear wavelet filter for parametric imaging with dynamic PET. *IEEE Transactions on Medical Imaging*, 22:289–301.

Urbanc, B., Cruz, L., Buldyrev, S. V., Havlin, S., Hyman, B. T., and Stanley, H. E. (1999a). Dynamic feedback in an aggregation-disaggregation model. *Physical Review E*, 60:2120–2126.

Urbanc, B., Cruz, L., Buldyrev, S. V., Havlin, S., Irizarry, M. C., Stanley, H. E., and Hyman, B. T. (1999b). Dynamics of plaque formation in Alzheimer's disease. *Biophysical Journal*, 76:1330–1334.

van Cappellen van Walsum, A. M., Pijnenburg, Y. A., Berendse, H. W., van Dijk, B. W., Knol, D. L., Scheltens, P., and Stam, C. (2003). A neural complexity measure applied to MEG data in Alzheimer's disease. *Clinical Neurophysiology*, 114:1034–1040.

van Putten, M. J. M. and Stam, C. J. (2001). Application of a neural complexity measure to multichannel EEG. *Physics Letters A*, 281:131–141.

Viergever, M. A., Maintz, J. B. A., Niessen, W. J., Noordmans, H. J., Pluim, J. P. W., Stokking, R., and Vincken, K. L. (2001). Registration, segmentation, and visualization of multimodal brain images. *Computerized Medical Imaging and Graphics*, 25:147–151.

Villa, A. E., Tetko, I. V., Dutoit, P., and Vantini, G. (2000). Non-linear cortico-cortical interactions modulated by cholinergic afferences from the rat basal forebrain. *Biosystems*, 58:219–228.

Wackermann, J. (1999). Towards a quantitative characterization of functional states of the brain: From the non-linear methodology to the global linear description. *International Journal of Psychophysiology*, 34:65–80.

Wackermann, J., Lehmann, D., Dvorak, I., and Michel, C. M. (1993). Global dimensional complexity of multi-channel EEG indicates change of human brain functional state after a single dose of a nootropic drug. *Electroencephalography and Clinical Neurophysiology*, 86:193–198.

Wada, Y., Nanbu, Y., Jiang, Z., Koshino, Y., Yamaguchi, N., and Hashimoto, T. (1997). Electroencephalographic abnormalities in patients with presenile dementia of the Alzheimer type: Quantitative analysis at rest and during photic stimulation. *Biological Psychiatry*, 41:217–225.

Waldman, A. D. and Rai, G. S. (2003). The relationship between cognitive impairment and in vivo metabolite ratios in patients with clinical Alzheimer's disease and vascular dementia: A proton magnetic resonance spectroscopy study. *Neuroradiology*, 45:507–512.

Wang, X. J. and Buzsaki, G. (1996). Gamma oscillation by synaptic inhibition in a hippocampal interneuronal network model. *Journal of Neuroscience*, 16:6402–6413.

Warkentin, S., Ohlsson, M., Wollmer, P., Edenbrandt, L., and Minthon, L. (2004). Regional cerebral blood flow in Alzheimer's disease: Classification and analysis of heterogeneity. *Dementia and Geriatric Cognitive Disorders*, 17:207–214.

Weir, B. (1965). The morphology of the spike-wave complex. *Electroencephalography and Clinical Neurophysiology*, 19:284–290.

Westbrook, G. L. (2000). Seizures and epilepsy. In Kandel, E. R., Schwartz, J. H., and Jessel, T. M., editors, *Principles of Neural Science*, pages 910–935. McGraw-Hill, New York, 4th edition.

Wickerhauser, M. V. (1994). *Adapted Wavelet Analysis from Theory to Software*. A. K. Peters, Wellesley, MA.

Williams, G. P. (1997). *Chaos Theory Tamed*. National Academy Press, Washington, DC.

Wolf, A., Swift, J. B., Swinney, H. L., and Vastano, J. A. (1985). Determining Lyapunov exponents from a time series. *Physica D*, 16(1):285–317.

Woyshville, M. J. and Calabrese, J. R. (1994). Quantification of occipital EEG changes in Alzheimer's disease utilizing a new metric: The fractal dimension. *Biological Psychiatry*, 35:381–387.

Xanthakos, S., Krishnan, K. R., Kim, D. M., and Charles, H. C. (1996). Magnetic resonance imaging of Alzheimer's disease. *Progress in Neuropsychopharmacology and Biological Psychiatry*, 20:597–626.

Xin, J. and Embrechts, M. J. (2001). Supervised learning with spiking neural networks. In *Proceedings of the International Joint Conference on Neural Networks*, volume 3, pages 1772–1777, Washington, DC.

Yagyu, T., Wackermann, J., Shigeta, M., Jelic, V., Kinoshita, T., Kochi, K., Julin, P., Almkvist, O., Wahlund, L. O., Kondakor, I., and Lehmann, D. (1997). Global dimensional complexity of multichannel EEG in mild Alzheimer's disease and age-matched cohorts. *Dementia and Geriatric Cognitive Disorders*, 8:343–347.

Zhang, B., Xu, S., and Li, Y. (2007). Delay-dependent robust exponential stability for uncertain recurrent neural networks with time-varying delays. *International Journal of Neural Systems*, 17(3):207–218.

Zhang, X. S., Roy, R. J., and Jensen, E. W. (2001). EEG complexity as a measure of depth of anesthesia for patients. *IEEE Transactions on Biomedical Engineering*, 48(12):1424–1433.

Zhou, Z. and Adeli, H. (2003). Time-frequency signal analysis of earthquake records using Mexican hat wavelets. *Computer-Aided Civil and Infrastructure Engineering*, 18(5):379–389.

Zhukov, L., Weinstein, D., and Johnson, C. (2000). Independent component analysis for EEG source localization. *Engineering in Medicine and Biology Magazine*, 19(3):87–96.

Index